ISMAEL SADOU

Insects Associated With Rice Cultivation

ISMAEL SADOU

Insects Associated With Rice Cultivation

In The Extreme North Sudano-sahelian Zone

ScienciaScripts

Imprint

Cover image: www.ingimage.com

This book is a translation from the original published under ISBN 978-3-639-60669-0.

Publisher:
Sciencia Scripts
is a trademark of
Dodo Books Indian Ocean Ltd. and OmniScriptum S.R.L publishing group

120 High Road, East Finchley, London, N2 9ED, United Kingdom
Str. Armeneasca 28/1, office 1, Chisinau MD-2012, Republic of Moldova, Europe
Managing Directors: Ieva Konstantinova, Victoria Ursu
info@omniscriptum.com

Printed at: see last page
ISBN: 978-620-8-51778-6

INSECT PESTS OF RICE IN THE SUDANO-SAHELIAN ZONE

CONTENTS

INTRODUCTION

When it comes to insects, rice is an extremely important host plant in terms of nutrition and reproduction. All parts of the plant are attacked, from the roots to the developing seeds. Insects that attack the seeds feed on the developing seeds. Defoliators feed on the leaves. Borer insects feed on the leaves and stems. Borers feed inside the stems and some species feed on the roots (Heinrichs, 1994). Worldwide, there are around 800 species of insect that cause damage to plants and stored seeds (Heinrichs and Barrion, 2004).

1. THE RICE PLANT

Rice (Oryza sativa L.) is the staple food of more than half the world's population (FAO, 2013).According to the United States Department of Agriculture USDA) (2014), records were set in both global rice production at 465.6 million and consumption at 468.7 million in 2013 (USDA, 2014).According to the FAO (2013), world rice consumption is expected to increase between 2012 and 2022 by just over 1% per year, compared with 1.7% for the period from 1990 to 2010. Average per capita consumption is expected to rise slightly compared with the reference period, to 58.2 kg per person (FAO, 2013).In Africa, increased production can be achieved both increasing the area under cultivation and by improving yields. In Cameroon, national rice production is estimated at 311,674 tonnes of paddy grown on 44,000 ha, covering only 38% of needs (FAO, 2013). As a result, imports have risen from 21,000 tonnes in 1975 to 728,443 tonnes in 2017, despite the country's potential, hence the need to increase production to reduce particularly expensive imports (FAO, 2017).Production by the Société d'Expansion et de Modernisation de la Riziculture de Yagoua (SEMRY) covers an area of 11,518 hectares, accounting for more than half of irrigated land and providing at least 90% of national production (FAO, 2013).African rice growing is subject to a number of constraints that seriously hamper production. There are two types of obstacle: abiotic obstacles (soil and climate) and biotic obstacles: insects, weeds, diseases, birds, nematodes and rodents (Heinrichs et al., 2004).Insect pests of rice constitute one of the main groups of biological agents limiting rice production in Cameroon (Nacro, 2014). The main insect pest species commonly found on rice fall into several orders (WARDA, 2016).However, according to Nacro (2014), four (4) orders stand out as being the most important. These are Lepidoptera, Isoptera, Coleoptera and Diptera, which cause considerable damage to rice fields (Vaissayre, 2007).In Cameroon, pest attacks are one of the reasons for low yields. Yield losses can be as high as 30% for rainfed rice and total for lowland rice (Dakouo and Nacro, 2012). The use of chemical insecticides was favoured to deal with these large and diverse groups of pests (Vaissayre et al., 2006). As soon as they appeared on the scene, synthetic insecticides attracted attention because of their efficacy and the immediate benefits they brought (Crétenet et al., 2015). Generally speaking, the advent of pesticides made it possible to curb the action of pests, and they were seen as a panacea before coming to a halt (Guèye, 2012).

1.1. General information on rice

1.1.1. Origin and geographical distribution

The origins of indigenous African rice, Oryza glaberrima, or floating rice, date back around 3,500 years to the Niger delta, where it was widely cultivated (Bouafia, 2002). Its cultivation has been replaced by that of Oryza sativa, imported from Asia, and is now essentially limited to northern Nigeria and Sierra Leone (FAO, 2004). Varieties of Oryza sativa from Guyana and Sri Lanka were introduced into Nigeria around 1890, but the precise dates and origins of the first introductions into Africa are not known (Polaszek et al., 2000). There are many cultivated varieties of Oryza sativa, used according to the type of rice cultivation practised (Amuli, 2015). Nigeria is the only African country south of the Sahara that currently produces a significant amount of rice, with most other countries relying mainly on imports (FAO, 2004). Madagascar is another major rice producer (Polaszek et al., 2000). However, it is increasingly used as a source of traits of agronomic interest in rice breeding programmes for Africa (FAO, 2004).

The domestication of Oryza sativa from Oryza rufipogon in India and China dates back more than 8,000 years (FAO, 2004). It is thought to have arrived in Japan in the ^{1st} century. Malay navigators introduced rice from Indonesia to Madagascar around the 4th century (Amuli, 2015). Europeans introduced Asian rice to Africa from the 15th century and later to America (Amuli, 2015).

Oryza sativa comes in a wide variety of forms. These forms have been classified into two subspecies, indica and japonica (Amuli, 2015). Initially based on morphophysiological characteristics and crossing behaviour, this classification has been confirmed by biochemical and molecular tools for analysing genetic variability (Bouafia, 2002).

Cultivated rices belong either to the species Oryza sativa, the most widespread, which varieties originating in Asia, or to the species Oryza glaberrima, cultivated in West Africa (Delseny, 2016). All varieties of Oryza sativa are thought to derive from the wild species Oryza fatua, which grows wild in India, Malaysia, Java and the Indochinese peninsula (Bouafia, 2002).

Its introduction into Central Africa is attributed to the Arabs (Amuli, 2015). It was not introduced to the Congo until 1840 by the Arabs. But the real expansion in the country took place during the rice-growing campaigns of 1935-1955, during which several varieties selected by the Institut National pour l'Etude Agronomique au Congo (INEAC) were disseminated (Bouman et al., 2011).

Rice is a monocotyledonous plant in the Gramineae or Poaceae family (Delseny, 2016). Cultivated rice belongs to two botanical species of the genus Oryza (Oryza sativa and Oryza glaberrima). The most widely cultivated species is Oryza sativa L, which is diploid, self-pollinating and native to South and South-East Asia (Second, 1982). It is thought to have been domesticated from Oryza rufipogon in India and China over 10,000 years ago (De Datta, 1981). Europeans introduced Asian rice to Africa from the fifteenth century and later to America. The species Oryza glaberrima or African rice or Casamance rice originated in Central Africa and has been domesticated in West Africa for 3,500 years (Dobelmainn, 2010).

A third rice species resulting from a cross between Oryza sativa L (Asian rice) and Oryza glaberrima (African rice) is grown in Africa and the rest of the "New African Rice" or NERICA (WARDA, 2006). Its germplasm offers real hope for improving the productivity, profitability and sustainability of rice growing in Africa (WARDA, 2006).

1.1.2. Rice classification

There are many different ways of classifying rice and there are several classification systems based on different criteria.Rice can be classified as follows (Vaughan et al., 2003):
Kingdom : Plantae

Sub-region: Tracheobionta **Division:** Magnoliophyta **Class:** Liliopsida
Subclass: Commelinidae

Order : Cyperales

Family: Poaceae (grasses)

Genus: Oryza L., 1753

Cultivated rice belongs to the genus Oryza, in the grass family and order Poales (Schalbroeck, 2001). The Oryza genus contains more than thirty species, two of which are currently cultivated, namely Oryza sativa L. and Oryza glaberrima Steud (Amuli, 2015). The species Oryza sativa L. originated in India and China, and its cultivation spread Asia, then Greece and Rome, later Madagascar, Africa and finally America (Schalbroeck, 2001).

According to botanical classification, the species Oryza sativa L. comprises three subspecies: Japonica, Javanica and Indica (WARDA, 2014). It can be distinguished from O. glaberrima by its short, truncated ligule. The species Oryza glaberrima adapts to extremely varied hydromorphic soils (Amuli, 2015).

1.1.3. Biology the rice plant

The life cycle of annual rice consists of several stages (Gaufichion et al., 2010). During its growth, rice goes through three essential phases. A vegetative phase, from germination to the initiation of floral primordials; a reproductive phase, from this initiation to pollination; and finally a grain maturation phase. It should be noted that the duration of the vegetative phase varies significantly between species under the same growing conditions, while that of the reproductive phase remains more or less constant whatever the variety (WARDA, 2014).

1.1.4. Plant morphology

Rice is an annual grass (poaceae) with a stem that can reach 2 m in height, or even more in the case of floating and semi-floating varieties (WARDA, 2014). The base of the stem emits a bud that gives rise to a secondary stem, or talle (Bouafia, 2002). This talle, thanks to its buds, will in turn give rise tertiary, quaternary, etc. stems. It is all these tillers, a single plant, that make up the rice tuft. The inflorescence is a branching panicle of varying width and density, 20 to 40 cm long. The spikelets have a single hermaphrodite fertile flower with 6 stamens. The glumellae, which are often aristate ("bearded" rice), completely envelop the grain once it has formed. The fruit is a caryopsis. Together with surrounding glumes and glumellae, it forms what is known as paddy rice (Bouafia, 2002). When ripe, its colour varies from light straw to deep purple and golden hues. The caryopsis consists of the integuments and the albumen (WARDA, 2014). The integuments can be variously coloured: reddish brown, grey, purple. The albumen is more or less translucent, depending on its content of amylose. In glutinous varieties, the albumen is opaque, white and chalky; its amylose content is close to zero. The weight of 1,000 paddy grains varies from 20 to 45g (Abdoulaye et al., 2013). The leaf consists of a blade, a sheath, two auricles (auricles) and a ligule. The blade, which is generally hairy, is 30 to 50 cm long and 1.0 to 1.5 cm wide (Abdoulaye et al., 2013). The sheath completely encloses the internode. The auricles are located on either side of the base of the leaf blade, and the ligule is a membranous, triangular, often bifid structure, 10 to 45 cm long, which extends the upper part of the sheath (Abdoulaye et al., 2013). The last leaf is the panicle leaf, sometimes called the flag leaf or standard; its blade is wider than that of the other leaves (Schalbroeck, 2001).

Each germinated grain gives rise to a tuft of up to thirty tillers at the vegetative stage (Abdoulaye et al., 2013). A limited number of these tillers, up to around fifteen, produce panicles. The height of the plant at maturity varies from 0.60 m to over 2 m depending on the variety, and can be up to 5 m for floating varieties (WARDA, 2014).

1.1.5. Rice physiology

According to Schalbroeck (2001), the best soil conditions are necessary for good rice growth.

1.1.5.1. The floor

Irrigated rice prefers heavy soils with low percolation water losses. Loose loam to clay-loam soil is best suited to rice cultivation (CIRAD-GRET, 2009).The optimum pH varies between 6.0 and 7.0, but irrigated rice can tolerate pH values of 4.5 to 8.5, because after a rice field is submerged, the pH that was acidic increases and the pH that was basic decreases by about two units (Schalbroeck, 2001). For dry cultivation, rice requires rich, loose soil with good water retention capacity and an optimum pH of between 6 and 7.

1.1.5.2. Temperature

Below 12 - 13°C, rice germination is impossible and some varieties require a minimum temperature of 18°C (Schalbroeck, 2001). Seedlings can recover after transplanting at an average daily temperature of around 13 to 15°C (Wopereis et al., 2008).

Table 2 below shows the air temperatures required for growing rice.

Table 1: Air temperatures required for rice cultivation, according to Wopereis et al.al. (2008)

Development stage	Air temperature (oc)		
	Minimum	Optimum	Maximum
Germination	14-16	30-35	42
Tallage	16-18	28-30	40
Flowering	22	27-29	40
Maturation	/	25	40

1.1.5.3. Water requirements

Irrigated rice growing with a permanent water table requires a water table of 1,000 to 1,500 mm, taking into account losses in the canals. On filtering soil, this amount of water can be doubled (WARDA, 2006).In rainfed rice farming, a well-distributed rainfall of 800 to 1000 mm is sufficient for a vegetative cycle of 4 to 5 months (Romains, 2001).Depending on climatic conditions and the length of the vegetative cycle, water requirements in terms of potential evapotranspiration are between 400 and 800 mm (Songbo, 2011).

1.1.5.4. Light

Sunshine plays an important role in rice growth and yield by encouraging tillering and increasing the number of spikelets and seed weight. A minimum of 400 hours of sunshine during the first two months of cultivation, including 220 to 240 hours during the last month, is necessary to obtain a high yield (Dobelmainn, 2010).

Seed dormancy (5 to 8 weeks) is controlled by genetic factors and growing conditions during the growth and ripening processes (WARDA, 2014). Japonica varieties do not have dormancy. In indica rice, the The longer the cycle, the more pronounced the dormancy, which does not normally occur in short-cycle varieties (Abdoulaye et al., 2013). In some varieties, dormancy is linked to photoperiodism, with long days accentuating dormancy. Dormancy can be lifted by various treatments, the effectiveness of which varies from variety to variety: by temperature (keeping dry grains at around 45°C for one to two weeks), by moisture in the presence of oxygen (e.g. top layer of soil) and by soaking in sodium hypochlorite or nitric acid solutions (WARDA, 2014). Dormancy is a handicap when you want to grow the same variety twice a year and use the harvest from the first season to sow the second. Dormancy has no effect on the longevity of seeds, whose germination capacity drops rapidly after one year under normal storage conditions.

Rice is a short-day plant whose sensitivity to photoperiod differs from one variety to another. Short-cycle varieties tend to be insensitive or only slightly sensitive to photoperiod, whereas long-cycle varieties are generally sensitive (WARDA, 2014). In addition, indica varieties are generally more sensitive than japonica varieties. Tropical rices, mostly of the indica type, flower more quickly for light durations of 10 to 12 hours, whereas japonica rices tolerate days of 14 to 15 hours (WARDA, 2014). Insensitive varieties can be sown in any season and lend themselves well to double cropping (Schalbroeck, 2001).

1.1.6. Rice vegetative cycle

The vegetative cycle of rice can be divided into three phases and five phenological stages (Lacharme, 2001):

- The vegetative phase, which comprises two stages: the seedling stage and the tillering stage.
- The reproductive phase also comprises two stages: the panicular initiation stage, followed

by bolting, and the heading, flowering plus fertilisation stage.

• The ripening phase, which has just one stage: the seed ripening stage. This phase begins just after fertilisation of the ovule and is characterised by progressive physiological changes in the fertilised ovule, leading to a mature grain.

The vegetative cycle varies between 90 and 160 days depending on the variety and sensitivity to day length. (Schalbroeck, 2001).

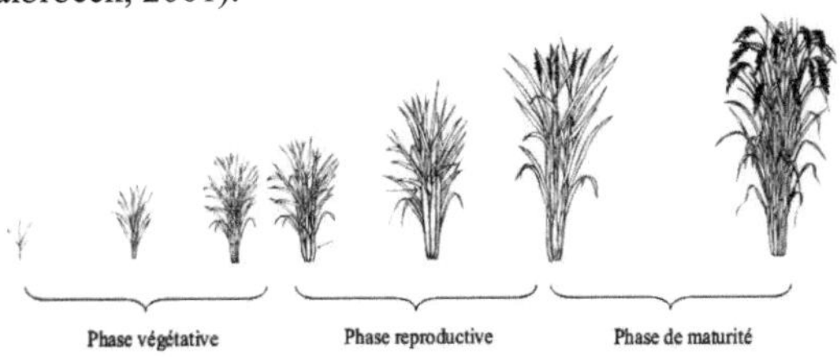

Figure 1: The three main phases in rice development, according to Wopereis et al., 2008

1.1.7. Rice phenological data

Rice has four clearly differentiated phenological stages (Table 1). These phenological stages sometimes have a very strong influence on the fauna of insects living to the detriment of rice, on the modalities of their attacks and subsequently on the crop losses they cause (WARDA, 2014).

Table 2: Characteristics of rice phenological stages

Phenological stages	Features
Germination (Nursery)	Seed germination and emergence; simultaneous emergence coleoptile and coleorrhysis, then simultaneous growth of both elements
Elongation (Tallage)	Successive emission of different tillers, according to the tillering process of a grass (the final number of tillers is a varietal characteristic).
Mounting	The panicle emerges after the last internode immediately below its insertion has elongated.
Flowering (Epiaison)	Opening of the spikelets, emergence of the stamens and fertilisation of the ovules. This very fleeting stage corresponds anthesis and ends with the closure of the spikelets.
Maturation of seeds	Progressive physiological changes in the fertilised egg leading to gradually towards the ripe grain

1.1.8. Rice production constraints

There are a number of obstacles to rice production in Africa. There are two types of obstacle: abiotic obstacles (soil and climate) and biotic obstacles: weeds, diseases, birds, nematodes, rodents and insects (Heinrichs et al., 2004).

2. INSECT PESTS ASSOCIATED WITH RICE CULTIVATION

Insect pests of rice constitute one of the main groups of biological agents limiting rice production in Cameroon (Nacro, 2014). The main insect pest species commonly found on rice fall into several orders (WARDA, 2016).However, according to Nacro (2014), four (4) orders stand out as being the most important. These are Lepidoptera, Isoptera, Coleoptera and Diptera.Of these groups of insect pests, stem-boring Lepidoptera and Diptera remain the key enemies on SEMRY's rice-growing perimeters (Dakouo, et al., 2018). Among the Lepidoptera borers, Chilo zacconius Bleszynski (Photo 1); C. diffusilineus de Joannis; Maliarpha separatelia Ragonot (Photo 2); Sesamia calamistis Hampson (Photo 3) (WARDA, 2016). Among Diptera, the diopside fly; Diopsis sp. (Photo 5) and the rice midge (Orseolia oryzivora Harris et Gagne) (Photo 6) are the main pest species (WARDA, 2016).

2.1. Order Lepidoptera

a) Chilo spp

According to Bonzi (2017), the genus Chilo (Photo 1) is the most important group of rice stem borers. The two species commonly found on rice, Chilo zacconius Bleszynski (syn. Procera africanaï and C. diffusilineus de Joannis) are morphologically very similar, with larvae appearing identical (Meijerman et al., 2014). These two species can be distinguished from each other by the shape of the male and female genetalia (Maes, 2016).According to Bordat et al (2010), unlike other borers, Chilo zacconius does not normally stay in dried rice stubble during intercropping. As it has no winter diapause, this borer finds refuge on weedy grasses in places that are kept moist, where it can reproduce. According to Maes (2016), the female Chilo zacconius lays a batch of 80 overlapping eggs on the leaf blades of rice plants. The number of eggs varies from 10 to 135 and the incubation period is 4 to 6 days (WARDA, 2010). In Chilo zacconius, oviposition can last up to 7 days, but production in the last two days is lower (Coquart et al., 2011).After a short feeding phase on the epidermal tissues, the young larva penetrates the rice stem, digs a gallery and completes its development before pupating (Betbeder, 2012).

Photo 1: Chilo zacconius (source: SADOU I., 2019)

b) Maliarpha separatella Ragonot

According to Maes (2016), Maliarpha separatella Ragonot (Photo 2) is the most important stem borer in the Phycitinae subfamily. It is at the tillering stage that the female lays her eggs (6 to 7 batches of around 40 eggs glued to the upper surface of the leaf by a cement which, as it dries, causes a characteristic pinching of the leaf blade completely enveloping the eggs (Breniere, 2013). The incubation period is 7 to 10 days. The larval stage lasts from 40 to 70 days and the larva goes through 5 to 7 stages (WARDA, 2016). The pupal period lasts from 32 to 65 days. There is rarely more than one larva per stem (Maes, 2016). Pupation occurs 30 to 50 days after hatching in the lower internode. The larva develops entirely in one (1) or two (2) internodes, is unable to migrate from one tillers to another and cannot survive on a plant without an internode (Akinsola et al., 2014). The larva and mature pupa are 22 mm and 15 mm long respectively.

Photo 2: Maliarpha separatella Ragonot (source: SADOU I., 2019)

c) Sesamia calamistis Hampson

The adult (Photo 3) lays 350 eggs deposited in batches of 10 to 40 without any particular alignment in the leaf sheaths that protect them (Nacro, 2017). After hatching, the larvae penetrate directly into the stem.Holloway (2016), reports that the duration of the larval stage (30 to 60 days) depends on climatic conditions. The larva undergoes 5 to 6 moults, and may attack several young stems in succession. At the pupal stage, the generation takes place in the stem and rarely between the leaf sheath and the stem. The larval and pupal periods last between 10 and 14 days and 28 and 35 days respectively (Nacro, 2017).

Photo 3: Sesamia calamistis Hampson (source: SADOU I., 2019)

2.1.1. Damage

Stem borer damage is caused by the development of larvae in the stem. According to Bacye (2016), the extent and nature of the damage caused larvae feeding in the stem depends on the phenological stage of the plant. For his part, Ingram (2016) indicates that the severity of damage varies according to the season, the type of rice cultivation, the zone and depends on cultivation practices and the type of stem borers present. Two types of damage are often observed (Akinsola et al., 2014; Nacro, 2014; Toure, 2017).An early attack during the vegetative phase leading to destruction of the central bud and drying of the central leaf, characterised by the "dead heart" symptom (Toure, 2017). When the attack occurs during the reproductive phase (flowering), the larvae feed inside the flower stalk, which dries out to form an entirely white panicle, with the spikelets emptied of their seeds; this is the "white panicle" symptom. The stem may then break. According to Bordat et al. (2010), the stem attacked by Chilo zacconius is generally completely destroyed, either at tillering or at heading. Maliarpha separatella causes total stem detachment at the internodes (Akinsola et al., 2014). Moreover, stem borer attack is less damaging when it occurs at tillering, because the plant can compensate by producing other tillers (Heinrichs, 2008).

2.2. Order Diptera

This order groups together several species including the diopsid flies Diopsis spp. and the rice midge Orseolia oryzivora Harris & Gagné.

2.2.1. Diopsis spp

According to Heinrichs (2008), Diopsis apicalis (Photo 4) and Diopsis thoracica (Photo 5) are the two (2) most important species commonly found in rice fields. Diopsis thoracica is the most abundant and widespread species in humid areas and infrequent in dry areas (Breniere, 2013). The Diopside fly can be distinguished by the characteristic shape of the adult, which has two very long peduncles on the head bearing the eyes on a bulge at the tip. The biology of the two species is more or less identical, but Diopsis apicalis differs in that it is smaller in size and wing tips are black (Breniere, 2013). The female deposits the eggs between the leaf sheath and the stem (Nacro, 2014). Incubation lasts 2 to 3 days. The larval stage lasts 4 to 5 weeks and pupation 10 to 22 days. The total length of the cycle is 60 days. The larva feeds inside young tillers. According to Nacro (2014), Diopsides attacks occur at the beginning of tillering, and in its lifetime, a single larva can destroy up to ten neighbouring stems. Damage caused by Diopsides is limited to young stems during the tillering period. It is manifested by delays in bolting and a reduction in the number of fruiting stems (Okocha, et al., 2017). Each affected tillers shows the symptom of "dead heart" and the tillers are lost. According to Akinsola et al (2014), the damage is similar to that of stem-boring Lepidoptera.

Photo 4: Diopsis apicalis (source: SADOU I., 2019)

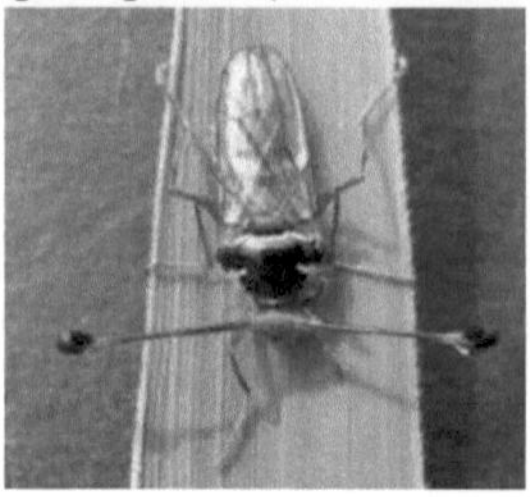

Photo 5: Diopsis thoracica (source: SADOU I., 2019)

2.2.2. African rice midge, Orseolia oryzivora Harris & Gagne

The African rice beetle is a tiny insect resembling a mosquito.According to Dakouo et al (2016), the Asian species Orseolia oryzae Wood Mason (Photo: 6) was long considered to be the only one present on the Asian and African continents. These two species show morphological distinctions in the larval, pupal and adult stages (Harris et al., 2015). Orseolia oryzivora has emerged as the most important insect pest of rice (Dakouo et al., 2018). The females settle on the leaves of young seedlings, where they deposit their eggs (200 to 200 eggs deposited in batches of 2 to 10) (Harris et al., 2015). After hatching, the larvae penetrate the leaf sheath to the apical point. The larvae only develop on tillers before bolting (Breniere, 2013). Two to three or more generations may be observed per crop and development from egg to adult takes 26 to 36 days (WARDA, 2010).Damage is caused by the activity of the larva, which feeds inside the infested young talus. It gnaws at the apical point (bud), causing tissue elongation and the formation of tubular structures called galls that are silvery in colour and resemble onion leaves (hence their respective names of "silver shoots", "leaves" or "onion tubes") (Perera & Fernando, 1970 cited by Harris et al., 2015). As a result, the formation of tillers always results in the loss of tillers. According to Dakouo et al (2018), the extent of damage varies depending on the year and cropping practices.

Photo 6: Orseolia oryzae Wood Mason (source: SADOU I., 2019)

Orseolia aryzivora is native to Africa. It was first discovered in 1947 in southern Sudan, but is now found in more than 20 countries south of the Sahara (Harris and Gagné, 1982). Prior to 1982, it was considered to be the same species that exists in Asia (Orseolia aryzae), but Harris and Gagné (1982) highlighted morphological differences between the two species.It has been present in Cameroon, Niger and Togo since 1955. In West Africa, it is found in all agro-ecological zones except the Sahel. It appeared in Nigeria in 1960, Zambia in 1984, Tanzania in 1991 and Sierra Leone in 1994 (Anonymous 1984, Alam et al 1985, Singh 1995).

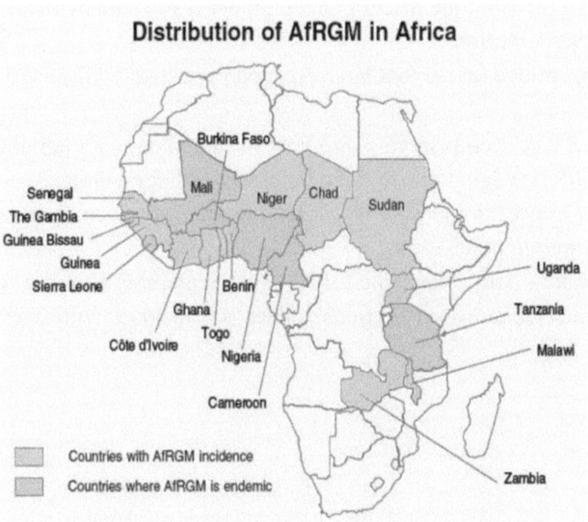

Photo7: Distribution AfRGM in Africa

- **Description**

❖ Adult

Adults have one pair of wings and three pairs of legs. The male is about 3 mm long, with a yellowish-brown body, while the female is about 3.5 mm long, with a light red abdomen. The male has a long, curved antenna, while that of the female is short (Bouchard et al 1992).

❖ Biology

The biology of these insects has been extensively studied by researchers (Panda and Mohanty 1970, Perera and Fernando 1970, Hidaka et al 1974, Kalode et al 1977). The male generally emerges very early in the morning before the female. As soon as the female emerges, mating takes place immediately. The male can mate with 2 to 4 females, whereas the female can only mate during her life cycle. The sex ratio is usually 1 male to 3 females (1:3). Environmental conditions can sometimes change this ratio. Often it is 1:6 or even 1:0.5. Sain and Kalode (1988) show that the offspring of Orseolia oryzivora are unisexual.

❖ Annual cycle

The annual cycle of the African rice midge can be summarised as follows:

• At the start of wet season, the midge migrates from alternate hosts to the newly established rice crop. Its populations are maintained on these hosts at a low level throughout the wet season.
• At the same time, in rice, thanks to particularly favourable climatic and biotic conditions, its populations are growing until the rice reaches maturity.
• At the end of the wet season, the insect's development is reduced by parasitism and unfavourable weather conditions.
• After harvest, some midge larvae that have escaped parasitism will survive on rice regrowth and other hosts.
• The introduction of a new crop serving as a host for the midge will not allow the population of this insect to proliferate again due to unfavourable weather conditions.
• We therefore have to wait for another wet season to see a spectacular recovery in the development of the midge population.

January February March April May June July August September October November December dry season rice rainy season rice residues rice residues wild rice: Oryza longistaminata

Caption:

High population level

Low population level

Source : INERA

❖ Lifecycle

Adult females live between 48 and 78 hours, while males only live between 23 and 30 hours. The complete life cycle (egg - egg) takes between 24 and 36 days under normal conditions of temperature and humidity. The egg stage lasts 4 to 6 days, the first larval stage 5 to 7 days, the second larval stage 5 to 7 days, the third larval stage 6 to 9 days and the pupal stage 4 to 7 days.

Table 3: Life cycle of Orseolia aryzivora under different temperature conditions

Conditions	Incubation period (day)	1st larval stage (day)	2th larval stage (day)	3th larval stage (day)	Stage pupal (day)	Total duration (days)
Normal (24-35oC)	4-6	5-7 /	5-7	6-9 /	4-7 /	24-36
Controlled (23oC)	3	7-9	/	6-8	6-8	23
Wind (16-36oC)	5-6		2-5			25-36

Relative humidity 85-90

Source: Sain (1988), Mohapatra et al (1996)

❖ Ecology

Rice plants remain the first refuges for Orseolia aryzivora in both rainfed and irrigated rice-growing ecosystems. Recent work has shown that in the dry season certain grasses (Leersia hexandra and Echinochloacrus- galli) and wild rice (Oryza nivara, O. barthii and O. rufipogon) serve as alternative hosts (Israel et al 1970, Prakasa Rao 1983).

Table 4: alternative hosts for Orseolia aryzivora

Hosts	Level of infection	Reference
Leersia hexandra Echinochloa crus-galli Wild rice Oryza nivara O. barthii O. rufipogon O. perennis O. glaberrima	to 10 galls/plant 1.8/plant to 10% galls/plant	Natarajan et al (1989) Sain (1988) Israel et al (1963) Hidaka et al (1983) Srivastava (1986)

Life cycle on rice= 20 to 25 days; on L. hexandra 25 to 30 days; on wild rice 25 to 30 days.

Source: Sain (1988), Mohapatra et al (1996)

❖ Laying and hatching

The eggs laid by the fertilised female just one day after mating are ovoid in shape. She can lay up to 125 eggs, which are deposited separately on the leaf blade or other parts of the leaf. They hatch 4 to 6 days after being laid. The individuals resulting from this hatching are all either males or females (Bergh et al 1990). Most males hatch within 2 to 3 hours in the evening. 12 to 15 hours later, the eggs with female individuals also hatch over a period of 2 to 3 hours.

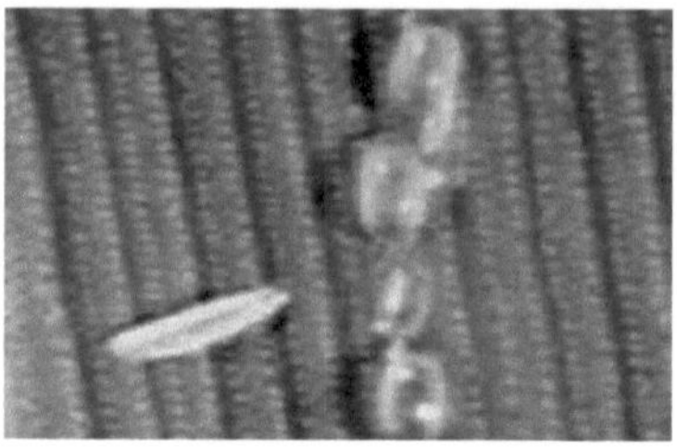

Photo 8: AfRGM eggs

- AfRGM larva and pupa

According to Mamaev (1968), midge larvae generally have two feeding mechanisms: a mechanism based on the type of food (they appear as sucking biters) and a mechanism based on the extra-oral digestion of the host plant (they appear here as secretaries of substances) from their salivary gland. They use specialised mouthparts to inject saliva into the plant, leaving its mark (Hatchett et al 1990, Grover 1995).

They start feeding from the leaf blade to the base of the plant. They live inside the stem, making production almost impossible. The migration from the leaf blade to the base of the stem, which only lasts between 12 and 15 hours, results in a high mortality rate (23-57 %) in the latter (McColloch and Yuasa 1917).

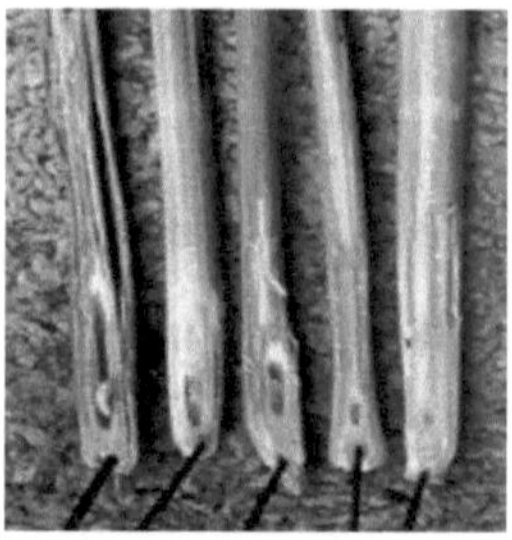

Photo 9: AfRGM larvae

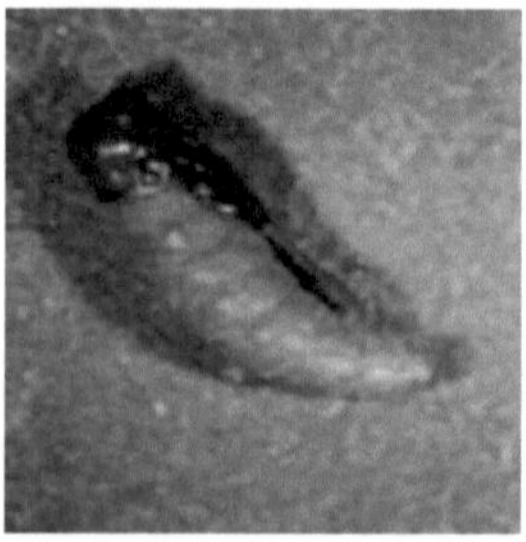

Photo 10 : AfRGM pupa inside stems

❖ Damage

The feeding site of Orseolia oryzivora depends on the growth stage of the plant. On young plants, the larvae of this insect feed at the base of the lower part of the leaves while infecting them (appearance of the gall). If the infection is severe, this can lead to the death of the plant. The plants then remain thin, with slightly white, curled leaves like onions. At the junction, they feed inside the stem approximately 0.5 cm above the node. At this stage, the disease spreads rapidly throughout the field, with the plant eventually becoming thin and dwarfed.

White hearts

Photo 11: White heart and rice gall in the field

Table 5: Areas affected by African rice gall in different countries

Country	Infected land	References
India	30-70 %	Mathur and Rajamani (1984)
Thailand		Hidaka et al (1996)
Indonesia	60 %	Hidaka and Budiyanto (1984)
China Lao		Di- yun and Cheng (1996)
PDR Sri	10 %	Khamhung (1996)
Lanka		Kudagamage et al (1988)
Vietnam	0.9 million ha	Hung and Anh (1996)
Africa		Ukwungwu et al (1989)
	Weaning attack in 1993- 1994 40	
	2000 - 5000 ha	
	45-80 %	

Source : K.C Mathur and K. Krishnaiah

❖ Control methods

The larval stage is the most devastating stage of the Orseolia oryzivora cycle. Its larvae feed inside the stem, making chemical control difficult. Nevertheless, the use of systemic insecticides such as carbofuran gives average results. At present, the only appropriate response seems to be the use of resistant varieties (IITA Annual Report and Research Update 1987/88). It should be noted that the use of biological control is also possible. Several researchers have observed natural enemies of this pest: Platygaster spp, which attack both eggs and larvae. They are the most abundant (90%). We also have Neanastatus sp (Hidaka et al 1974, Prakasa Rao 1983).

Table 6: Most common natural enemies of Orseolia aryzivora

Natural enemies	Infected stage
Parasitoids Platygaster oryzae P. foersteri Neanastatus grallarius Obtusiclava (Propicroscytus) oryzae Eurytoma sp Predators Ophionea indica Paederus fuscipes Amblyseius imbricatus	 Egg/larva Egg/larva Pupa Larva/pupa Larva/pupa Pupe Pupe egg

Source: Sain (1988), Mohapatra et al (1996)

Annual cycle of parasitoids of Orseolia oryzivora
January February March April May June July August September October November December rainy season rice wild rice: Oryza longistaminata
Caption:
High AfRGM parasitoid population level
Average AfRGM parasitoid population level
Low AfRGM parasitoid population level
Source : INERA

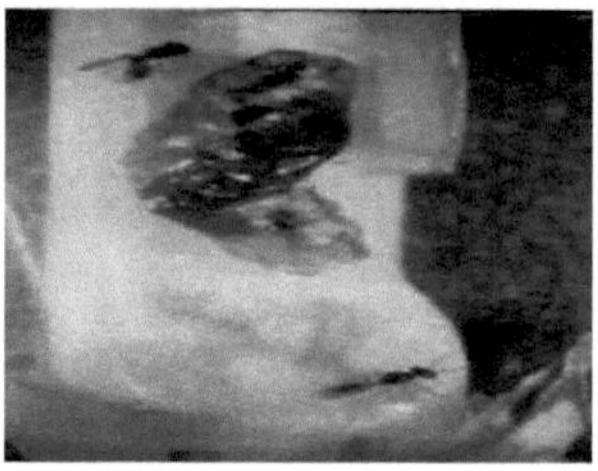

Photo 12: Platygaster sp

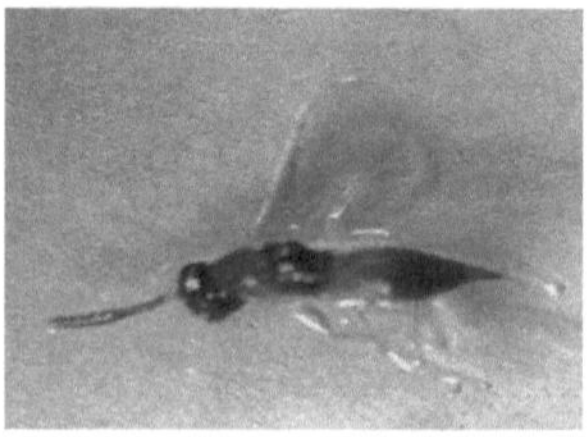

Photo 13: Aprostocetus sp

Egg parasitoids Parasitoids of pupae

2.3. Orders of Hemiptera

2.3.1. Leaf biting Hemiptera Families: Cicadellidae, Pentatomidae **Damage**

Hemiptera (aphids and cicadellidae), like Heteroptera (Pentatomidae), have an effect on plants through the toxicity of their saliva, resulting weakening of the plant, and their ability to cause disease. sap samples and, lastly, by possible inoculation of various diseases (viruses, mycoplasmas, bacteria) through punctures (Bourgoin et al., 2004).In addition, the sugar content of Hemiptera excrement provides a favourable environment for the development of various fungi. Cicadellidae and Pentatomidae mainly attack vegetative organs. When they occur in large numbers, Hemiptera stings cause the foliage to dry out and become covered in tiny brown spots (Swzedo et al., 2004).This damage is rare in Africa, despite the presence of many species of Hemiptera. In Asia, the economic importance of these pests has increased considerably over last decade. The hemipterans Nephotettix nigropictus (Photo 14) and Nilaparvata lugens Stal (Photo 15) have been recognised as vectors of serious viruses: "Tungro", "Yellow dwarf" and "Grassy Stunt Virus", which seem to be spreading with the development of highly productive, tall-growing varieties (Swzedo et al., 2004). There is currently no need to combat leafhoppers in Africa. The problem would only take on an economic dimension if Asian viruses succeeded in establishing themselves. In that case, control would essentially be based on varietal resistance to rice viruses.Chemical control of the vectors is only a transitional solution of dubious long-term effectiveness (appearance of insecticide-resistant strains) (Sampong et al., 1979).

Photo 14: Nephotettix nigropictus (source: SADOU I., 2019)

Photo 15: Nilaparvata lugens Stal (source: SADOU I., 2019)

2.3.2. Grain peckers

Family: Pentatomidae

- Scotinophara spp (Photo: 16)

Photo 16: Scotinophara spp (source: SADOU I., 2019)

Family: Coreidae

- Leptocorisa spp. (Photo: 17)

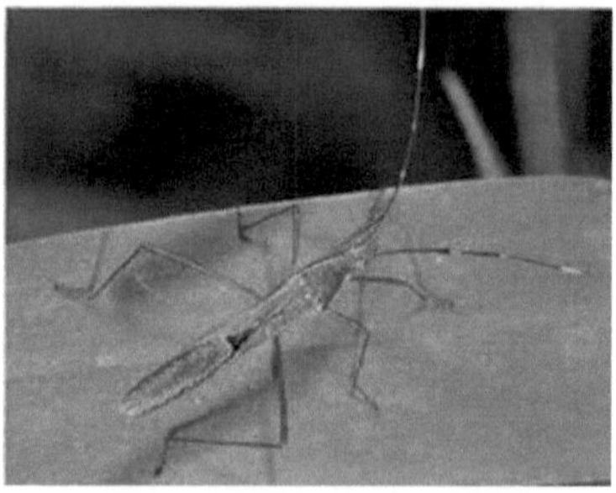

Photo 17: Leptocorisa spp. (Source: SADOU I., 2019)

- Stenocoris spp. (Photo: 18)

Photo 18: Stenocoris spp. (Source: SADOU I., 2019)

- Mirperus spp. (Photo: 19)

Photo 197: Mirperus spp. (Source: SADOU I., 2019)

- Oulema oryza (Photo: 20)

Photo 20: Oulema oryza (Source: SADOU I., 2019) Damage :

The grain may be completely or partially emptied. On rare occasions, large outbreaks of pentatomid beetles can occur, concentrating on isolated rice fields and causing very significant losses.In general, weight loss remains low. On the other hand, quality is affected by the presence of grains which, after milling and bleaching, have more or less extensive black spots. This defect cannot be eliminated by sorting, and even a small proportion of black grains seriously reduces the commercial value of the rice.Since the sale of paddy at the production level does not generally take quality into account, these pests are not controlled. However, the insect is easy to be destroyed with most current insecticides. We can assume that only a large rice-growing group that is well organised for quality intensive rice-growing could act at the right time and at the right price.

3. GROUPING OF INSECT PESTS OF RICE ACCORDING TO THEIR DAMAGE OR ATTACKS ON PLANTS

The diversity of their mouthparts, their diets and their ability to adapt to almost any environment make insects the most formidable and devastating of all crop pests (Dakouo et al., 2012). Rice cultivation is therefore faced with a number of problems caused by insect pests, which affect almost every part of the plant from the beginning to the end of the production cycle.According to Bacye (2016), around 4098 species of insect pests have been inventoried for rice crops alone. Most of these insect pests are Lepidoptera, Diptera, Coleoptera and Hemiptera, and some of them are chewers, chewer-lickers, suckers and others are sucking biters. According to Akinsola et al (2014), depending on their damage or attacks on plants, they can be grouped into five main groups which are:

- Stem borers ;
- Defoliators ;
- Biting and sucking ;
- The root cutters ;
- Stock pests.

3.1. Stem drills

According to Akinsola et al (2014), there are around fifty species divided into two orders: Lepidoptera and Diptera. The most important species in the order Lepidoptera are: Chilo spp, Maliarpha separatella, Sesamia calamistis, Nymphula depunctalis, Eldana saccharina, Scirpophaga spp. The Diptera order includes Diopsis spp and Orseolia oryzivora. The latter species causes rice midge. Stem borers are the most important pests, infesting rice plants from the seedling stage through to maturity. There are two main types of damage (Defoer et al., 2004):Damage caused to young plants at the beginning and during tillering. Here, the caterpillars penetrate the leaf sheaths and the base of the young stem; the damage can lead to the death of the stem.Damage caused from the panicle. The young caterpillars (especially of the ^{2nd} generation) gather a few centimetres below the panicle inside the flower stalk. This dries out completely, leaving an entirely white or dried-out panicle. Apart from these two highly visible cases, any damage to the stem is also harmful. When the insect attacks a plant at an advanced stage of development, the old caterpillar lodges in the lower parts of the stem and can reduce or stop the panicle's food supply. Damage results in the drying out of one or more branches, leading to a reduction in the number of harvestable grains. For these attacks by stem borers, order of Lepidoptera (Sesamia sp...) and Diptera (Diopsis thoracica West...) are the most represented (Wopereis et al., 2008).

Photo 21: Nymphula depunctalis (Source: SADOU I., 2019)

Photo 22: Eldana saccharina (Source: SADOU I., 2019)

Photo 23: Scirpophaga spp. (Source: SADOU I., 2019)

3.2. Defoliators

According to Betbeder (2012), the damage caused by leaf destroyers generally involves removing fragments from the tips of the leaves. In other cases, the leaves may be cut off completely. This damage is often of little economic importance when the plant is sufficiently developed. However, at the start of the crop and up to the middle of tillering, infestations can affect a large part of the plant, leading to irreparable destruction (Wopereis et al., 2008). The order most commonly found in this type of attack is Coleoptera and Orthoptera, while Lepidoptera includes Nymphula depunctalis, which is the most common species for this type of attack (Betbeder, 2012). It is a small white butterfly measuring 10 to 12 mm, which lays its eggs in tight rows along the leaves. As soon as it is born, the larva climbs onto a leaf and shears it off, leaving a flap that serves as an attachment (Betbeder, 2012). The leaf fragment curls up but doesn't swell and closes with a few silk films, finishes cutting off the rest of the leaf fragment, then drops into the water. It swims towards a rice stalk on which climbs and feeds. This allows them to move easily from one stalk to another. Drying up for at least three days is an effective means of control. Other defoliating species include : Cnaphalocrocis medinalis (Photo 24), Marasmia trapezalis (Photo 25), Diacrisia scortilla (Photo 26), Parnara spp (Photo 27) (Heinrichs et al., 2004).

Photo 24: Cnaphalocrocis medinalis (Source: SADOU I., 2019)

Photo 25: Marasmia trapezalis (Source: SADOU I., 2019)

Photo 25: Diacrisia scortilla (Source: SADOU I., 2019)

Photo 26: Parnara spp. (Source: SADOU I., 2019)

3.3. Picking and sucking

Biting and sucking insects can attack leaves, stems or seeds (Breniere, 2013). They are sometimes numerous and can cause significant damage in the event of a large outbreak. This category includes bugs, whiteflies, aphids, Cicadellidae and mites, which cause damage mainly during the ripening phase, at the milky and pasty stages. The most represented order in this group is the Hemiptera (Wopereis et al., 2008).

3.4. Root cutters

Root cutters are insects, most of which develop partially or totally in the soil (Breniere, 2013). Some are found exclusively in low-lying Others only occur in plateau conditions (Wopereis et al., 2008).

- **The rice moth**

These insects belong to the same order as locusts (Breniere, 2013). Although they are present in all ecosystems, crickets are more abundant in uplands and their damage is greatest at field edges.

- **Termites**

They can be found in low-lying areas as well as on plateaux (Betbeder, 2012). Their presence is always favoured by the absence of water. This is why flooding plots of land is the best way to combat these insects. They feed on the roots and underground parts of the plant, causing the leaves to turn yellow and dry out. The species most commonly found in rice fields are Microtermes, Macrotermes and Trinervitermes spp (Betbeder, 2012).

- **Water weevil**

This insect lives exclusively in the shallows and muddy waters of rice fields. The larva is the harmful stage that feeds exclusively on rice roots (Betbeder, 2012).

3.5. Stock pests

Stored rice is attacked by a wide range of insects whose identity and cycle differ from those encountered in the field (Sawadogo, 2017). Most of these are polyphagous insects that attack not only the rice but also other stored grains. Stored grain insects fall into two groups depending on the mode of infestation. Some start attacking crops in the field and are introduced into the granary or shop through the harvest. This is the case for weevils and alucites. Others only attack during storage; they come either from inside or outside the shop (Heinrichs et al., 2004).

4. ENVIRONMENTAL FACTORS CONDUCIVE TO THE DEVELOPMENT OF INSECT PESTS

Biotic factors such as entomophages and entomopathogens and abiotic factors such as rain, relative humidity and temperature exert regulatory effects on insect pest populations (Dakouo et al., 2018).

4.1. Abiotic factors

Various ecological factors favouring the development of the rice plant greatly benefit insect pests (Maes, 2016). Temperature, rainfall and relative humidity are factors that can affect the distribution, development, survival, behaviour, reproduction and dynamics of rice insect pests. According to ADRAO (2010), a temperature of 25°C, humidity of 70 to 75% and a photoperiod of 18 hours during the day and 6 hours at night encourage insect proliferation. Abundant rainfall and temperatures between 25 and 30°C are favourable for the development of rice midge (Shin Foon quoted by Nacro, 2017). Biswas et al (2017) report that water stress affects plant vigour, which becomes susceptible to insect attack. Insect activity can be influenced by certain cultivation practices. For example, very close spacings and high doses of nitrogen (Soraja et al., 2016), periodic irrigation (Rao, 2015) and early or late sowing increase the incidence of midge attacks (Rai et al., 2013). Crop associations on sorghum (sorghum-cowpea) reduce pest damage to crops (Zongo et al., 2016). Savary et al (2017) establish a relationship between cultivation practices, insect attacks and yield reductions.

4.2. Biotic factors

Host plants and natural enemies are factors that influence insect population dynamics (Rai et al., 2013). Two parasitoids, Platygaster diplosisae Risbec and Aptostocetus procera associated with African rice blast have been reported (Dakouo et al., 2018; Nacro, 2017). Cumulative parasitism rates of 70% of these two parasitoids have been recorded.

Photo 27: Platygaster diplosisae Risbec (Source: SADOU I., 2019)

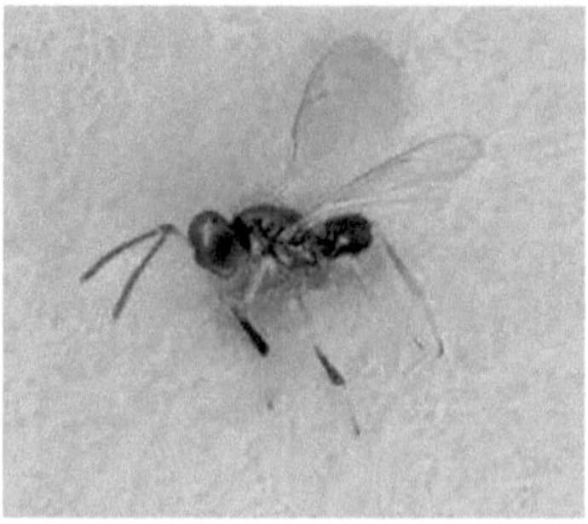

Photo 28: Aptostocetus procera (Source: SADOU I., 2019)

5. METHODS FOR CONTROLLING INSECT PESTS OF RICE

Nowadays, several control methods exist to enable better control of insect pests of rice (IRRI, 2015).

5.1. The natural enemies of insects

The natural enemies of insects are entomophages, organisms that feed on insects. Entomophagous insects include parasitoids, which develop in or on their host, eventually killing it, and predators, which pursue their prey or capture it by stalking or trapping (Cloutier, 2012).

5.1.1. Parasitoids

A parasitoid can be defined as: "an organism that develops on or in another organism, its host, derives its sustenance from it and kills it as a direct or indirect result of its development" (Boivin, 2018).According to this definition, parasitoids can be insects, nematodes, fungi, bacteria, protists and viruses (Eggleton et al., 2016). However, the majority of known parasitoids are insects (Boivin, 2016).The 87,000 species of parasitoid insects listed are divided into six orders: Hymenoptera (67,000), Diptera (15,), Coleoptera (4,000), Neuroptera (50), Lepidoptera(10) and Trichoptera (1) (Boivin, 2016, 2018). The two main orders in which parasitoids used in biological control are found are Hymenoptera and Diptera (Byers, et al., 2015).The Hymenoptera families most commonly used for biological control are the Braconidae, Ichneumonidae, Eulophidae, Pteromalidae, Encyrtidae and Aphelinidae. Diptera are mainly represented by the family Tachinidae (van Driesche et al., 2016).

5.1.1.1. Parasitoid insect hosts

The impressive diversity of parasitoid insects is reflected, among other things, by their wide variety of hosts (Vinson, 2014). Although most parasitoids attack insects, some species of parasitoids, especially diptera, choose spiders, chilopods and even vertebrates as hosts (Boivin, 2016).

In addition, parasitoids can be found at all stages of insect development: egg, larva, pupa and adult. Some species can even parasitise one stage of development and emerge at the next. For example, Holcothorax testaceipes is a Hymenoptera of the Encyrtidae family that lays its eggs in those of Gracillariidae (a lepidopteran), develops in the egg and caterpillar and finally emerges from the latter (van Driesche et al., 2016). Generally, there are no differences between the hosts used for the development of male and female parasitoids (Boivin, 2016). However, in heteronomous parasitoids, represented mainly by the family Aphelinidae, males and females develop differently (Strand, 2016).

Indeed, in some species, males and females can develop from the same species, but the former (ectoparasitoids) develop on the host while the latter (endoparasitoids) develop inside the host (Strand, 2016).In other species, the females are parasitoids of the primary host (usually a herbivore), but the males are parasitoids of the female eggs of their own or another parasitoid species (secondary host) (Strand, 2016).

This phenomenon is often referred to as adelphoparasitism. Finally, in some cases, both males and females are primary parasitoids, but the host of the males is a different species to that of the females (Godfray, 2014).

The particular adaptation of species with adelphoparasitism makes their use in biological control more restrictive. In these species, such as Coccophagoides similis and C. utilis, unmated females produce only males, while mated females produce both females and males (Strand, 2016).

Thus, since the males of these species parasitise the eggs of the females, the unmated females must be released after the mated females. In this way, it is possible to control self-parasitism and male production in these species (Sullivan 2017).

5.1.1.2. Parasitoid reproduction

Generally, parasitoid insects lay their eggs inside or outside the host (Jones et al., 2014). In the first case, they are called endoparasitoids and in the second, ectoparasitoids.

To develop inside their host's body, endoparasitoids have to deal with the host's immune defences (Hassell et al., 2016).

When parasitized, several species secrete haemocytes that encapsulate the parasitoid egg and prevent its development. Some species of endoparasitoids avoid encapsulation, either by depositing their eggs in a site protected from the host's haemocytes, or by transmitting a virus that affects the host's immune system during oviposition (Greathead, 2016).

By laying their eggs outside the host, ectoparasitoids have the advantage of avoiding the host's immune defences, but are more likely to be injured or dislodged. To overcome these disadvantages, many ectoparasitoids inject a venom that partially or totally paralyses the host (Strand 1986).

Other species lay their eggs in the surrounding environment, in which case it is the larva that has to find its host (Cloutier et al., 2012). For example, female Aleochara bilineata, a beetle in the Staphylinidae family, lay their eggs in the soil near the host plant. Once they emerge, the first instar larvae seek out and parasitise the pupae of certain Diptera species.

Being both a parasitoid (larva) and a predator (adult), A. bilineata helps stabilise populations of Delia radicum, a major pest of cruciferous crops (Royer et al., 2018).

In fact, D. radicum is controlled not only by A. bilineata larvae, but also by adults, which consume large quantities of fly eggs and larvae. An adult can consume up to 23.8 fly eggs or larvae per day (Royer et al., 2018).

5.1.1.3. Hyperparasitoids

Apart from insecticides, predators and pathogenic micro-organisms, parasitoids have to contend with during their short lives (Sullivan, 2017). Parasitoids are themselves subject to parasitism by other parasitoids (Boivin, 2016).

These are known as hyperparasitoids or secondary parasitoids. Hyperparasitism found in 17 Hymenoptera and a few cases in Diptera and Coleoptera (Sullivan, 2017).

There are two categories of hyperparasitoids: obligate and facultative. In the first case, hyperparasitoids develop solely in or on a parasitoid, while in the second, hyperparasitoids develop from a parasitoid. primary host, generally a herbivore, or a parasitoid attacking this herbivore (Sullivan, 2017). In the latter case, the choice between the primary or secondary host will depend on health of the host and whether or not the host is already parasitised (Jones et al., 2014).

For example, the hyperparasitoid Pachycrepoides vindemiae is a parasitoid of several Drosophila species. The latter can be parasitised by other parasitoid species such as Asobara tabida and Leptopilina heterotoma (Hassell et al., 2016).

When P. vindemiae encounters a Drosophila already parasitized by another species, it can lay eggs and develop as a facultative hyperparasitoid of the young parasitoids already existing in the original host (Godfray, 2014).

The choice of whether or not to parasitise depends on the stage of development of the immature parasitoids. If the primary host is not parasitized, it will lay eggs in it and the larvae will develop as primary parasitoids (Godfray, 2014).

There are a few known species of parasitoids that are known as tertiary parasitoids, i.e. they attack hyperparasitoids (Hassell et al., 2016). Given the detrimental role that hyperparasitoids can play on populations of beneficial parasitoids, precautions are generally taken to avoid introducing hyperparasitoids from the parasitoids that are to be introduced into a biological control programme (Sullivan, 2017).

5.1.2. Predators

Given their more generalist nature, insect predators have a much greater taxonomic variety than parasitoids (Holling, 2016). However, the arthropods group together the most interesting entomophagous predators, particularly the polyphagous insect species (Sunderland et al., 1983). Predators act in a simple and direct way: the prey is captured and generally eliminated without delay (Jackson, 2016). A predator can eliminate several prey during its lifetime, its voracity being a useful indicator of its potential for repression (Holling, 2016). Furthermore, predation requires less specialisation than parasitism, resulting in lower specificity than in the case of parasitoids (Holling, 2016). A polyphagous predator uses several prey species and the importance of each varies according to its relative availability (Jackson, 2016).However, no generalist predator is universal, with all showing clear preferences or abilities towards certain types, sizes or sets of similar ecological prey (Hanski et al., 2017).In addition, the prey used by a predator can vary with its age or size, with younger individuals having to make do with smaller prey that is easier to capture. The majority of predators actively pursue their prey and capture it by (Coccinellidae, predators of various Hemiptera) or by leaping (non-specialised predatory spiders) (Holling, 2015).

5.1.2.1. The concept of predation strategy

The quantity of food that a predator can appropriate, digest and convert into energy is limited, either by the abundance of prey, its accessibility or the time required to capture and assimilate it (Abrams, 2017). The organism must be able to allocate limited food and energy resources to its growth, maintenance, fight against its enemies and reproduction (Dawkins et al., 2016).

A predation strategy must therefore be able to answer the following questions: (Holling, 2015):

- What prey are you looking for?

- Where and when to eat?

- How to capture prey effectively (trade-off between capture effort and nutritional value of the prey).

Predators will therefore tend to optimise their strategies in order to make the most of the prey they hunt. Natural selection therefore favours, among the various possible strategies, the one that gives the predator population the best possible fitness (selective value) (Tirard, 2012).

This solution, known as the optimal strategy, will depend on the constraints weighing on the predator, such as the accessibility of the prey, its capturability, and the defences anti-predationput in place by the prey (escape strategies to better evade the predator, self-preservation, etc.). making it inedible by secreting venom, for example, the ability to camouflage itself in its environment, etc.) (Curio, 2012).

5.1.2.2. Different predation strategies

Predators have different diets (Jackson, 2016). In fact, some predator species can be polyphagous, i.e. have a very varied diet (Curio, 2012). Others may be oligophagous, i.e. consume only a small variety of foods. Others may be monophagous, i.e. they feed exclusively on a single species of prey. There are different predation strategies.

5.2. Methods of combating insect pests through cultivation practices

These are cultivation practices that prevent the multiplication of harmful insects and at the same time favour rice. The most important cultivation practices are :

- The destruction of rice stumps and harvest residues, where larvae and/or nymphs can survive from one season to the next;

- Respecting and synchronising the cropping calendar to avoid the proliferation of insects;
- The use of healthy or treated seeds ;
- Cultivation and destruction of primary infection sites or host plants (weeds);
- Crop rotation wherever possible to break the insect cycle and prevent proliferation;
- Combined or intercropping crops to limit the spread of insects plants.

5.2.1. The general biological method

According to Wopereis et al (2008), the aim is to limit the density of pests through the action of natural enemies such as predators, parasites or pathogens. Two strategies can be distinguished: conservation and increasing natural enemies. Conservation simply means not destroying natural enemies through cultural practices or other interventions such as chemical control. Augmentation, on the other hand, consists of artificially increasing the quantity of natural enemies in the environment. This involves breeding and releasing natural enemies, which is very rare in Africa. What is used is natural conservation, which leads to a natural increase.

A typical example is the two midge parasitoids. Farmers should be encouraged to keep Paspalum scrobiculatum around their fields, as it is the preferred breeding ground for these two parasites. The more this plant is produced, the more the populations of natural enemies increase around the field, which naturally leads to a drop in midge populations (Nacro, 2017).

Furthermore, it has been proven that after weeding, if grass residues are piled up around the rice stems, this leads to an increase in spider populations, which hide there during the day and hunt nocturnal stem borers at night (Johnson, 2017).

These different methods are known as conservation and augmentation methods. Several types of cultivation practices are therefore conservation methods in biological control.

Other biological control solutions could be envisaged, in particular the promising use of highly specific sexual attractants (pheromones) which, once synthesised, could contribute to the destruction of pests by trapping or in combination with an insecticide (Hari, 2017).

This process is beginning to be developed in the case of certain lepidopterans of fruit crops. The pheromones of Chilo suppressalis and Chilo zacconius are known and synthetic components are currently being tested (Toure, 2017).

These 'products' have a promising future, particularly in the field of 'prediction'.

"This is because sex attractants can be used to detect very low insect populations, particularly at the start of outbreaks. However, pheromones are not currently available in Africa.

Biological control has the advantage of posing little danger to health or the environment. However, this type of control is often relatively complex to manage and requires substantial investment and a certain level of technical expertise.

5.2.2. The genetic method

This control method consists of using tolerant varieties; it is generally inexpensive for farmers and not very harmful to the environment.

Varieties can be distinguished according to their resistance and classified as low, medium and high resistance (Hari, 2017). There are two types of genetic resistance: vertical resistance, which is the resistance of varieties against certain insects, and horizontal resistance, which is resistance against most insects.

5.2.3. The integrated method

This control method involves using tolerant varieties (IRRI, 2015). It is generally inexpensive for farmers and not very harmful to the environment.

Varieties can be distinguished according to their resistance and classified as low, medium and high resistance (Hari, 2017). There are two types of genetic resistance: vertical resistance, which is the resistance of varieties against certain insects, and horizontal resistance, which is resistance against most insects.

5.2.4. The chemical method

Chemical control using insecticides has the advantage of generally being fairly effective, killing the insect in a short space of time (Insah, 2017). In addition, insecticides are often widely available in shops. However, the use of insecticides is fairly expensive and poses technical constraints.

Liquid sprays are not very precise and are less effective than granular treatments, which have the advantage of being applied by hand or with inexpensive equipment. Some farmers use furadan to prevent the proliferation of stem borers.

There is a danger associated with the frequent use of insecticides because some insects have the ability to adapt to insecticides and thus become resistant (Insah, 2017). What's more, most insecticides are non-specific and kill both harmful and beneficial insects. The use of insecticides therefore poses a great danger to the ecosystem and the environment, and this alternative should only be applied as a last resort, after all other existing options have been considered (FAO, 2018).

CONCLUSION

Insect pests of rice are a very serious threat to rice production (irrigated and rainfed) in the Far North region of Cameroon and cause yield losses. They have not received the attention they deserve as a major constraint to rice production in Africa.The studies enabled an inventory to be made of insect pests of rice in rice-growing ecosystems in the Far North region. These inventories made it possible to list several species of insects in the order of Hemiptera, Diptera, Coleoptera, Lepidoptera, Hymenoptera, Odonata, Orthoptera, and also the order Araneae, which are native to these ecosystems. The insects that cause most damage are Maliarpha separatella, Sesamia calamistis belonging to the order Lepidoptera, and Diopsis sp which is a Diptera. Maliarpha separatella and Sesamia calamistis are choppers in the larval stage (caterpillar); their larvae and Diopsis sp stem borers that caused an estimated 23.68% damage to the crop, resulting in a drop in yield. The most represented order is the Diptera with Physiphora clausa F, Culex robinotus, Glyphodera mantis, Diopsis sp, Microdon jahanna, Paragus dolichorus and Allongrata nasuta Macqueat and least represented order is the Hymenoptera with Bracon sp, Apanteles ruferierus. It can also be seen that the rainfed rice variety NERICA 3 is much more favoured (attacked) by insects than the irrigated rice variety IR 46.In addition to the presence of insect pests of rice, a number of species of natural enemies have been identified namely predators and parasitoids. Predators are much more numerous at mature stage. But it should be noted that spiders are the most important predators, so natural enemies help to control other insect pests. Among the natural enemies, the predatory species Tetragnatha juculator (Araneae: Tetragnatidae) and Lestes spp (Odonata: Lestidae) are present at all phenological stages. The presence of these predators is of great importance and helps a great deal in the natural regulation of insect pest and disease (vector) populations.Losses caused by insect pests on the rice crop were 49.98%. These results show the importance of pest control in increasing production. But it should be noted that apart from the losses caused by insect pests, granivorous birds, diseases and some rodents also cause damage to rice.

BIBLIOGRAPHICAL REFERENCES

Abdoul A., Hughes J. & Diallo A. (2001). Yellow mottle in rice. Economic importance, diagnosis and management strategies. WARDA Proceedings 1: 250 - 251.

Abdoulaye s, Abou, Togola, & Ali, Toure. 2013. Proposition pour une optimisation des performances a la riziculture en Afrique de l'ouest Senegal: 34pp.

Abo M. E., Alegbejo M. D., Sy A. A. & Sere Y. (2004). Retention and transmission of Rice yellow mottle virus (RYMV) by beetle vectors in Cote d'Ivoire. Revue de l'Association ivoirienne des sciences agronomiques 16 : 71 - 76.

Abrams P. A. (2017). The evolution of predator-prey interactions: theory and evidence, Annual Review of Ecology and Systematics, pp. 2-4.

Adégbidi, A. (1994). Cours de gestion des exploitations agricoles. FSA/UNB; p81.

Adégbidi, A. (2018). Socio-economic determinants of the adoption of new technologies: The case of maize in the department of Atacora in the Republic of Benin. Postgraduate doctoral thesis in economics (rural economics), CIRES, University Côte d'Ivoire, Abidjan, 169p.

Adégbola, P. and Midingoyi, S. (2004). Efficacité technico-économique des agriculteurs et rizicultrices du Centre-Benin. Paper presented at the NATIONAL 2004 scientific workshop.

Adégbola, P. & Sodjinou, E. (2017). Study of the rice sector in Benin. Final report; p 231.

ADRAO. (1990). WARDA Annual Report, 1990. 1: 2-36.

WARDA. (2000). Highlights of Activities. WARDA Annual Report, 2000. 1 : 1-12.

WARDA. (2010). Insect pests ofrice in west Africa WARDA, Bouaké (Côte - d'Ivoire). 73 p

WARDA. (2014). NERICA rice variety as a source of life in West Africa. Bouaké, Ivory Coast: 64pp.

WARDA. (2016). Annual Report. WARDA, Bouaké (Côte - d'Ivoire). 59 p. WARDA. (2017). New Rice for Africa. www.warda.org.

Agarwal P. C., Mortensen N. C. & Mathur B. S. (1994). Seed-borne diseases of rice and phytosanitary tests. WARDA. 80 p.

Ahoyo, R.N.A. (2018). Economics of production systems integrating rice cultivation in South Benin: potentialities, constraints and perspectives. PhD thesis, Peter Lang, 270 pages.

Akinsola E. & Agyen - Sampong M. (2014). The ecology, bionomics and control of rice stem - borer in West Africa. Insect Sei Applic. 5 (2): 69 - 77.

Akol A.M., Sithanantham S., Njagi P.G.N., Varela A., and Mueke J.M. (2002). Relative safety of sprays of two neem insecticides to Diadegma mollipla (Holmgren), a parasitoid of the diamondback moth: effects on adult longevity and foraging behavior. Crop Protect, 21:853- 859.

Aminatou B. (2009). Inventaire des phytoinsecticides pour la protection des grains au cours stockage contre les ravageurs dans la zone sahélienne (cas de l'Extreme Nord du Cameroun). Final report, Réseau Anafe/Raft-AHT. 32p.

Amuli (2015). Comparison of two intermittent water management techniques on the yield of five irrigated rice varieties at Kiringye in the Ruzizi plain, unpublished dissertation, UCB.

Anonymous (2010). Mémento de l'agronome, P.691 - 1254 éd. Quae, Montpellier - France.
Appert J. (1970). Maliarpha separatella (White borer of rice). Observations nouvelles et rappel des problèmes entomologiques du riz à Madagascar, Agronomie Tropicale 25: 329-362.

Arhent C., Cramer H. H., Mogk M. & Peshel H. (1983). Economic impact of crop losses. 10th International Congress of Plant Protection. November 10- 25, 1983. 1: 65 - 73.

Arraudeau M. (1998). Irrigated rice. Maisonneuve & Larousse, Paris. 656 p.
Arraudeau, M.A. and Vergara, B.S. (2018). A farmer's primer on growing upland rice. Los Baños, Philippines, IRRI. 95 pages.

Attikou, A. (2014). Suivi de la fertilité des sols des aménagements hydro-agricoles : Cas des périmètres rizicoles de Bonféba et Lata. Niamey, Office National des Aménagements Hydro-Agricultural. 39pp. Akintayo I., Cissé B., and L. D. Zadji. 2008. Guide pratique de la culture des NERICA de plateau. Cotonou, Benin: Africa Rice Center (WARDA), 36pp.

Awinakai J. (1998). Rapport de stage de découverte d'un milieu humain. University of Dschang. 42 p.

Awoderu V. A. (1991). The Rice Yellow Mottle virus situation in west Africa. Tropical pest management 37: 356-362.

Bacye. B. (2016). Contribution to the development of a rational control system for rice insect pests. Phytosanitary monitoring and experimentation on the rice-growing perimeter of the cou valley. Final dissertation, IDR Université de Ouagadougou. 76 p.

Bamba, B. (2015). Contribution to the evaluation of losses due to rice blast. Dissertation, University of Ouagadougou. 51 p.

Barker, R. & Herdt, R.W. (2016). Rainfed lowland rice as a research priority - an economist's view. In Rainfed lowland rice, pp. 3-50. Los Baños, Philippines, IRRI.

Beauvilain A. (2011). Notes on the cities of the northern province. Revue de Géographie du Cameroun, 2 (1): 25-32.
Beckage N.E. (1985). Endocrine interactions between endoparasite insects and their hosts. Annual Revue of Entomology 30: 371-413.

Bélanger A. & Musabyimana T. (2005). Neem against insects and diseases. In: Journée horticoles 2005. 4.
Bénech V., Quensière J., Vidy G., (2012). Hydrology and physicochemistry of North Cameroon floodplain waters. Cah. Orstom, ser. Hydrol, 19 (1): 15-36.

Bernard J.R. Philogène, Cathérine Regnault-Roger & Charles Vincent (2008). Biopesticides of plant origin: assessment and prospects. 2nd edition, Edition Tec & Doc, pp1-24.

Bessong à Beyeck, L. (2006). Potential trade between Cameroon and its border partners. Mémoire d'Ingénieur d'Application de la Statistique, option gestion. Institut Sous Régional de Statistique et d'Économie Appliquée (I.S.S.E.A).

Betbeder Matibet, M. (2012). Rearing of several species of the genus Chilo and some of their parasites for biological control of grass borers in Africa. Insect.Sci. Applie, 11 (5/5): 617-623.

Bhan, V.M. (2016). Relation between growth and yield characters in upland paddy. Riso, 15: 229-232.

Bhan, V.M. (2018). Influence of row spacing on upland paddy. Riso, 17(2): 149-154.
Blache J., Miton F. (2011). Première contribution à la connaissance de la pêche dans le bassin hydrographique Logone-Chari-lac Tchad. Paris, ORSTOM, Mémoire n° 4, 143 p.

Billon B., Bouchardeau A., Roche M., Rodier J. (2015). Monographie hydrologique du Logone. Paris, Orstom, 8 vols, 770 p.

Biofil (2008). The nim (or neem), the miracle tree.
Biscaldi R. (2014). Hydrology of the Logone-Chari-Chad water table. BRGM, Report 70-Yao-003.

Biswas, A. K., Islam, A., Nayek, B. & Choudhuri, M.A. (2017). Water - stress induced susceptibility to pests in rice plants (Oryza sativa). Indian Biologist 14 (1): 13 - 20.

Biswas K., Chattopadhyay I., Banerjee R.K., Bandyopahyay U. (2002). Biological activities and medicinal properties of neem (Azadirachta indica). Current Science, 82, 1336-1345.

Bodji J-L. & Yao K. A. (2017). Study of the efficacy of neem (Azadirachta indica Juss) and papaya (Carica papaya L.) plant extracts in the control of vegetable crop pests. Final dissertation for the Diplôme d'Ingénieur des Techniques Agricoles, Ecole supérieure d'Agronomie de Yamoussoukro, 56p.

Boeke, S. J. (2002). Traditional African plant products to protect stored cowpeas against insect damage; the battle against the beetle. Ph D thesis, Wageningen University, 151 p. + appendix Boeke, S.J. Boersma, M.G. Alink, G.M. VanLoon, J.J.A. Huis, A. Dicke, M. Rietjens,
I.M.C.M. (2004). Safety evaluation of neem (Azadirachta indica) derived pesticides, Journal of Ethnopharmacology 94 : 25-41.
Boivin, G. (2016). Evolution and diversity of parasitoid insects. Antennae. Special issue: 6-12.

Boivin, G. (2018). Entomophagous research: state of play. Annales de la Société Entomologique de France (N.S.) 35 (suppl.): 348-355.

Bonzi. (2017). Results of the first investigations on insect pests of rice in Upper Volta. Paper presented at the seminar of the Association for the Development of Rice Cultivation in West Africa (ADRAO) in Bobo - Dioulasso from 17 to 22 September 2017.

Bordat, D., Breniere, J. & Coquard, J. (2010). African grass borers: parasitism and husbandry techniques Agron. Troptc. 32 (4). 391-399.
Bouafia A, 2002, La rousse agricole: 755 pp.
Bouchardeau A. (2015). Monographie hydrologique du Logone. Paris, Orstom, 8 vols, 770 p.
Bouman, B. A. M. & Tuong P. (2011). Field water management to save water and increase it

productivity in irrigated rice agriculture water management 49 (1), 11-30.
Bourbonnais G. (1991). Guidelines for collecting insects and arthropods. http://ici.cegep-ste-foy.qc.ca/pdf, consulted on 10/10/2008.

Bourgoin Th., J. Swzedo & F. Lefèbvre, 2004. About Hemiptera phylogeny and Classification, :11-36. In: Fossil Planthoppers (Hemiptera: Fulgoromorpha) of the world.

Brenière J. H. (1983). Principal pests of rice in West Africa and their control. Association pour le Développement de la Riziculture de l'Afrique de l'Ouest. 2nd Ed. 87 p.

Breniere, (2013). Major rice pests in West Africa and their control. West African Rice Development Association, Monrovia, Liberia. 52 p.

Brevault Guibert H. & Naudin K. (2007). Preliminary studies of pest constraints to cotton seedlings in a direct seeding mulch-based system in Cameroon. Expl Agric (2009), volume 45, pp. 25-33. Cambridge University Press.
Bruneton J. (1999). Pharmacognosy. 3rd edition, Editions Tec & Doc - Lavoisier, Paris.
Buffington, M. L., & R. A. Redak (2008). A comparison of vacuum sampling versus sweep-netting for arthropod biodiversity measurements in California coastal sage scrub. Journal of Insect Conservation 2: 99-106.
Byers, J.R., Yu D.S. & Jones. J.W. (2015). Parasitism of the army cutworm, Euxoa auxiliaris (Grt.) (Lepidoptera: Noctuidae), by Copidosoma bakeri (Howard) (Hymenoptera: Encyrtidae) and effect on crop damage. The Canadian Entomologist 125 : 329-335.

CBLT. (2016). Studies of water resources in the Lake Chad Basin, with a view to a development programme. In: Groundwater resources in the Lake Chad Basin, LCBC, Cameroon, Niger, Nigeria, Chad, UNDP, Rome FAO.

Callahan, R. A., Holbrook, F. R. & Shaw. F. R. (2016). A comparison of sweeping and vacuum collecting certain insects affecting forage crops. Journal of Economic Entomology 59: 478-479.

Cessouma A. (2008). L'agriculture irriguée et perspective et l'économie d'eau au Mali 67pp.
CIRAD-GRET, (2009). MEMENTO de l'agronome, New edition, Ministère des affaires étrangères (France) Pp799-811.

Christian Seignobos & Olivier Iyébi-Mandjek, (2005). Atlas de la province Extrême-Nord Cameroun, Paris, IRD Éditions, 171 p

Coquard, J. & Breniere, (2011). Chilo zaccontus Biez rice borer in West Africa: Fecundity on artificial medium Agron. Tropic. 34 (2) : 175-179.

Côté S. (2004). Comment commencer sa collection d'insectes. Association des Entomologistes Amateur du Québec (AEAQ) 1: 25-23

Cloutier, C. and C. Cloutier. 2012. Biological control solutions for the suppression of insect and mite pests of crops. Pages 19-88 in: C. Vincent and D. Coderre (eds.), La lutte biologique. Gaëtan Morin, Boucherville.

CTA. (1995). Integrating biological control and host plant resistance. Proceedings of a CTA/IAR/IIBC seminar. Addis Ababa, Ethiopia, 9-14 October 1995. 383 p.

Colombani J., Olivry J.-C., & Kallel R., (2014). Exceptional erosion and solid transport phenomena in arid and semi-arid Africa. In: Proceedings of the Symposium of Harare, IAHS Publ. no. 144: 299-300.

Curio E., (2012). The ethology of predation, Springer Science & Business Media, vol. 7.

Dakouo, D. & Nacro, S. (2012). La lutte contre les insectes ravageurs du riz au Burkina : bilan et perspectives in Lutte intégrée contre les ennemis des cultures vivrières dans le Sahel INSAH (edit.), Séminaire international de lutte intégrée Bamako 4 - 9 janvier1990, pp170 - 186.

Dakouo, D., Nacro, S., & Sie, Ivi. (2016). Seasonal evolution of rice midge infestations, Orseolia oryzivora H. et G (Diptera, Midge) in southwest Burkina Faso. Insect Sei. Applic. 9 (4) pp 469-473.

Dakouo, D., Nacro, S. (2018). Development of a rational insect pest management system on SEMRY rice fields. Insect. Sc. Appl. 15 (5/6): 565- 570.

Dawkins R., Krebs J. R., (2016). Arms Races between and within Species, pp. 489-511; DOI: 10.1098/rspb.1979.0081. [1] [archive].

De Datta, S.K. & Feur, R. (2017). Soils on which upland rice is grown. In Major research in upland rice, pp. 27-39. Los Baños, Philippines, IRRI.

De Datta, S.K. & Vergara, B.S. (2017). Climate of upland rice regions. In Major research in upland rice, pp. 14-26. Los Baños, Philippines, IRRI.

Decraene P. (2017). Lettre de Maroua: l'ombre portée du Nigeria. Le Monde, 29 to 29 July 2014: 8.Defoer (2004). Facilitator's manual: insect pests of rice; 6 pages.

Defra. (2013). Evaluation of fully approved or provisionally approved products. Evaluation on Lambda-cyhalothrin Use as a public Hygiene insecticide (Food and environment protection act, 1985, part III) issue n°104, Department for Environment, Food and Rural Affairs (DEFRA) - Pesticide Safety Directorate.

De Groot, I. (1996). Protection of stored leguminous cereals. CTA. Agrodoc18: 3-45. Delseny M. (2016). Rice, a model in genomics. Biofutur 203, 30-35.

Delvare G. & Aberlenc P. 1989. Insects of Africa and Tropical America. Clé pour la reconnaissance des familles. 297 p.

Denham M. (2014). Voyages et découvertes dans le nord et les parties centrales de l'Afrique, exécutés pendant les années 1822, 1823 et 1824. Trad. Eyries et de la Renaudiere, Paris, A. Bertrand, 3 vols. 367 p., 379 p., 428 p.

Diaz, OL (2009). Climate change, poverty and transport in cities south of the Sahara. Université Cheick Anta Diop, 310p.

Dobelmainn J.P. (2010). Riziculture pratique. Riz pluvial. Techniques vivante. Agency for Cultural and Technical Cooperation. Presses universitaires de France, Paris; France.

Dubreuil P. (2015). Étude des inondes sur un petit bassin de la région de Maroua, Nord-Cameroun. Paris, ORSTOM, Annuaire hydrologique de la France d'Outre-Mer: 15-27.

Durand J. R. & Lévêque C. (2015). Aquatic flora and fauna of Sahelo-Sudanian Africa. Paris, Orstom, IDT 45, 484 p.

Eggleton, P. & Gaston K.J. (2016). Parasitoid species and assemblages: convenient definitions or misleading compromises? 59: 417-421.

Ekukolé G. & Ajayi O. (1995). Some observations on Scymnus (Scymnus) floralis (F.) and Ischiodon aegyptius (Wiedmann), predators of the cotton aphid in North of Cameroon. Journal of African Zoology 109: 93-97.

Engola-Oyep J. (1991). Assurer la survie des périmètres hydro rizicoles à l'heure de l'ajustement structurel. Cahier des Sciences Humaines 27: 53-63

WARDA Spirit. (2004). The Africa Rice Center. WARDA Journal 5: 2 - 4. FAO. (2000). Rice. FAO. Statistical Bulletin 1: 20-73.

FAO. (2001). Statistical Yearbook; www.fao.org. 31/03/2017. FAO. (2002). Statistical yearbook of production 56 :175-176.

FAO. (2004). Extraordinary Summit of Heads of State and Government of the African Union on Employment and Poverty Reduction in Africa. Situation of agriculture and rural populations in Africa.

FAO. (2011). The state of food insecurity in the world. How international price volatility affects the economy and food security of countries. Rome 62 p.

FAO. (2013). Rice. FAO. Statistical Bulletin 2: 2-10.

FAO. (2014). FAO. Statistical Bulletin 1: 24-54.

FAO. (2017). Improved rainfed rice farming systems, Rome.

FAO. (2018). Statistical Yearbook; www.fao.org.

Faye M. (2010). New fractionation process for Senegalese neem seed (Azadirachta indica A. Juss): production of a bio-pesticide oil and cake. Thesis, Département des Sciences des Agro-ressources, Université de Toulouse. 267p.+ appendix.

Fotsing E., (2009). Small savannah: An information system for the integrated analysis of land use change in the Far North of Cameroon. PhD thesis.

Fritsch P., (2013). Geological aspects of the floodplains of North Cameroon.

GauldI. D. & Bolton B. (1988). The Hymenoptera. British Museum Natural History Press 1: 331-332.

Gilburn A.S., Bunnefeld N., Vean Wilson J., Botham M.S., Brereton T.M., Fox R., & Goulson D. (2015). Are neonicotinoid insecticides driving declines of widespread butterflies? https://doi.org/10.7717/ peerj.1402.

Gnago J. A., Danho M., Atcham Agneroh T., Fofana I. K. & Kohou A.G., (2016). Efficacy of neem (Azadirachta indica) and papaya (Carica papaya) extracts in the control of insect pests of okra (Abelmoschus esculentus) and cabbage (Brassica oleracea) in Côte d'Ivoire. Int. J. Biol. Chem. Sci.4(4): 953-966, August 2010.

Godfray, H.C.J. (2014). Parasitoids. Behavioral and evolutionary ecology. Princeton University Press. New Jersey.

Goldstyn J. (2003). Capture methods 2. Montréal Insectarium. 54 p.

Goswami, N.N, Bapat, S.R., Leelavathi, C.R. & Singh, R.N. (2016). Potassium deficiency in rice and wheat in relation to soil types and fertility status. Indian Soc. Soil Sci. Bull, 10: 186-194.

Greathead, D.J. (2016). Parasitoids in classical biological control. Pages 290-318 in J. Waage and D. Greathead (eds.), Insect parasitoid. Academic Press, London.

Guigaz M. (2000). Mémento de l'agronome. CIRAD GRET, Paris. 1690 p.

Guillard J. (2015). Studies and work on fishing in the Cameroonian Logone Basin. Garoua, Archiv. de l'Inspection forestière du Nord.

Guiscafre J. (2016). Influence des aménagements anti-érosifs sur l'écoulement des mayos Kapsiki - Bassins versants de Mogodé. Yaounde, Orstom-Ircam, 40 p., graphs.

Gupta, P.C. & O'Toole, J.C. (2018). Upland rice - global perspective. Los Baños, Philippines, IRRI. p. 1; 63-68; 306-310.

Hamadoun A. & Traoré (1997). Status of rice mottle in Mali: the case of the Office du Niger. Pp 20-22 In: International Symposium Rice Yellow Mottle Virus (18-22 September 1997). M'bé/ Bouaké, Ivory Coast.

Hanks, J. (2006). Mitigation of human-elephant conflict in the Kavango-Zambezi Transfrontier Conservation Area, through Community Based Problem Animal Control, with particular refrence to the use of chilli peppers. Report prepared by Conservation International.

Hanski I, Hansson L & Henttonen H. (2017). Specialist predators, generalist predators and the microtine rodent cycle. Journal of animal ecology 60-1 353-367 doi:10.2307/5465.

Harris &Gagné. (1982). Description of the African Rice Gall midge with comparative notes on the Asian rice gall midge, O. oryzae (wood Mason) (Diptera: Cecidomydae). Bulletin entomology 72: 467-472.

Harris, K.M., & Gagne, RJ., (2015). Taxonomy of Asian and African rice gall midge. International Rice Research Newsletter, 7 (2): 8.

Hari K. P., (2017). Improved rainfed rice farming systems Food and Agriculture Organisation Rome, Italy, 1997. 172 p.

Hassell M.P. & Godfray H.C.J. (2016). The population biology of insect parasitoids. Pages 265-293 in M.J. Crawley (ed.), Natural enemies. The population biology of predators, parasites and diseases. Blackwell Scientific Publications, Boston.

Heinrichs E. A. (1983). Biology and management of rice insects. Ed. E.A. Heinrichs, New york. 779 p.

Heinrichs E. A. (1994). Rice. Pp 3-11 In: Biology and management of rice insects. Heinrichs ed. Wiley and Eastern Limited New Delhi, India.

Heinrichs E.A. &Alberto T. B. (2004). Biology, ecology, identification. Rice-Feeding Insects and Selected Natural Enemies in West Africa. Ed. P. Hettel, Bouaké. 450 p.

Heinrichs E. A., Barrion A.T. (2004). Rice- feeding insects and selected natural enemies in West- Africa. Biology, Ecology, Identification. International Rice Research Institute and WARDA, the Africa Rice Center, Bouaké. 80 p.

Heinrlchs, E.A., (2008). Rice West Africa in African cereal stem borer. Economic importance, taxonomy, natural enemies and control. Edited by A. Polaszek. Center for agriculture and biosciences international (London), pp: 49-57.

Hill D. S. (1983). Agricultural insect pests of the tropics and their control. LondonUniversity Press 2: 745-746.

Hirsch, R. (2018). African rice farming: importance and issues (rice and rice policies in West Africa and the PSI/CORAF zone). In Pour un développement durable de l'agriculture irriguée dans la zone Soudano-Sahélienne : Synthèse des résultats du Pôle régional de Recherche sur les systèmes irrigués (PSI/CORAF), Proceedings of the Seminar Dakar (Senegal) from 30 November to 3 December 1999, CIRAD/CF/CTA.

Holling C. S. (2015). The components of predation as revealed by a study of small mammal predation of the European Pine Sawfly, The Canadian Entomologist, vol. 91, pp. 293-320.

Holling C. S. (2016). Some Characteristics of Simple Types of Predation and Parasitism, The Canadian Entomologist no. 91, pp. 385-398. Doi : 10.4039/Ent91385-7.

Hubertus Kleeberg, (2015). Le neem, un agent de contrôle biologique à base d'extraits végétaux in Biopesticides d'origine végétale, 2nd edition, Editions Tec & Doc, p14.

Hugon Ph. (2014). Agricultural sectors and macroeconomic policies. In Benoit-Cattin M, Griffon M, Guillaumont P, éditions économie des politiques agricoles dans les pays en développement, les aspects macroéconomiques, Paris, revue française d'économie.

Huke R.E. & Huke E.H. (1990). Rice. Then and now. Manila, International Rice Research Institute 1: 1- 44.

IITA (International Institute of Tropical Agriculture). (1987). Annual report for 1987. IITA, Ibadan. 160 p.

IITA (International Institute of Tropical Agriculture). (2000). Annual Report and Research Update 2000. IITA, Ibadan. 145 p.

IITA (International Institute of Tropical Agriculture). (2013). Annual report and research update 2013. IITA, Ibadan. 192 p.

Ingram, W. R. (2016). Biological control of graminaceous stem-borers and legume podborers. Insect. Sei. Appl, 4 (1/2). Pp 205-209.

Insah, (2017). Integrated pest management of food crops. Second seminar on integrated pest management of food crops in the sahel. 392 P.

IOMC (Inter-Organization Programme for the Sound Management of Chemicals). (2002). Pesticide Risk Reduction Steering Group. Report of the organization for economic cooperation and development (OECD), Pesticide Risk Reduction Group Seminar on Minor Uses and Pesticide Risk Reduction. 18 p.

IRRI (International Rice Research Institute). (1997). Annual report. IRRI, Manila. 90 p.

IRRI (International Rice Research Institute). (2015). Problems in rice cultivation. Identification guide second ed. Manila Philippines. Pp :71-79.

IRRI. (2017). Rice facts. Los Baños, Philippines. Manila. 87 p.

IRRI. (2018). International upland rice genetic survey. Los Baños, Philippines. Manila. 45 p.

Isman M.B. (2017). Neem insecticides. Pesticide outlook, 8 p.

Isman M.B. (2018). Botanical insecticides, deterrents and repellents in modern agriculture and increasingly regulated world. Annu Rev Entomol, 50p.

Issoufou, A. & Issa, G. (2006). Détermination des surfaces et estimation de la production de riz hors aménagements hydro-agricole. Study report. Niamey, Programme d'Appui à la Filière Riz. 43 pp.

Jackson R. (2016). Aggressive mimicry, prey-specific predatory behaviour and predator-recognition in the predator-prey interactions of Portia fimbriata and Euryattus sp., jumping spiders from Queensland. 98 pp.

Javed N., Gowena S.R., El-Hassana S.A., Inam-ul-Haqa M., Shahinab F. & Pembroke B. (2008). Efficacy of neem (Azadirachta indica) formulations on biology of root-knot nematodes (Meloidogyne javanica) on tomato. Crop Protection, 27, 36-43.

Jeroen B., Koomen I., Joep.V. L de Jeude. & Oudejans.J. (2014). Pesticides: composition, use and risks. In serie Agrodok 29, 1ère édition. Editions Digigrafi & wageningen, The Netherlands. 124 p.

Johnson, D.E. (2017). Les adventices en riziculture en Afrique de J'Ouest, WARDA, Bouaké (Cote -d'Ivoire), PP: 1-2.

Jones, T.H., M.P. Hassell & R.M. May (2014). Population dynamics of host-parasitoid interactions. Pages 371-397 in B.A. Hawkins and W. Sheehan (eds.), Parasitoid Community Ecology. Oxford University Press, New York.

Karimi J., Haubruge E., Francis F. (2010). Development of entomotoxic molecules as control agents: illustration of some protein potential uses and limits of lectins (Review), Biotechnologie Agronomie.Société et Environnement. 14 (1) : 225-241.

Kaung V. T., John & Alam. (1985). Rice production in Africa: an over view, in Rice Improvement Eastern, central and southern Africa. Pp 7-27 In: Proceeding of the international

Rice workshop. International Rice Research Institute, Philippines.

Kessler S.C., Tiedeken E.J., Simcock K.L., Derveau S., Mitchell J., Softley S., Stout J.C. & Wright G.A. (2015). Bees prefer foods containing neonicotinoid pesticides. Nature 7: 74-76.

Klatt B.K., Rundlöf M., & Smith H.G. (2016). Maintaining the Restriction on Neonicotinoids in the European Union - Benefits and Risks to Bees and Pollination Services. Frontiers in Ecology and Evolution 4: 4. http://dx.dor.org/10.3389/fevo.2016.00004.

Kouassi N. K., N'Guessan P., Albar L., Fauquet C. M. & Brugidou C. (2005). Distribution and Characterization of Rice yellow mottle virus: A Threat to African Farmers. The American Phytopathological Society 89: 124-133.

Kosma P., Ambang Zachée, Ten Hoopen M. & Begoude D. (2013). Study of the susceptibility of plantain cultivars to Radopholus similis (Cobb) Thorne (nematode) and protection trials with neem seed formulations (Azadiracta indica A. Juss). 1re African Research Conference on Agriculture, Food and Nutrition. Yamoussoukro, Côte d'Ivoire, 4-6 June 2013. Les Presses Agronomiques de Gembloux, A.S.B.L. Pp: 511-524.

Kumar R. (1991). Insect pest management: the situation in African agriculture. CTA/Karthala, Wageningen, Paris, 310 p.

Kumar, A.R.V. Jayadevi, H.C. Ashoka, H.J. Chandrashekara, K. (2003). Azadirachtin use efficiency in commercial neem formulations. Journal of Current Science 84 (11): 1459-1464.

Lesueur F. (2006). Development of formulations based on Azadirachta indica neem extracts. A. Juss pour la protection de la pomme de terre (Solanum tuberosum L.) contre le Myzus persicae, un puceron colonisateur et vecteur de virus circulants et non Circulants. Master's thesis, Université de Laval- Québec, 139p. + appendices.

Maes, K. V.N. (2016). Pyraloidea, Carabidae, Pyralidae. in African cereal stem borer. Economic importance, taxonorny, natural enemies and control. Edited by A Polaszek. Center for agriculture and biosciences International. (London), pp: 87-98.

Mahapatra, I.C. & Srivastav, V.C. (2014). Management of rainfed upland rice: problems and prospects. Oryza, 21: 20-37.

Marco Salvioni (2014). Evaluation of rat (Rattus rattus) damage in rice field in Madagascar. Tropical Pest Management 37 (2) - 175-178.

Matsushima S. (2016). Crop Science in rice, theory of yield determination and its application. Fuji Publishing Co, Ltd, edit.3.

Meijerman, L. & Ulenberg, S.A. (2014). Larvae (Morphology). in African cereal stem borer. Economic importance, taxonomy, natural enemies and control. Edited by A Polaszek. Center for Agriculture and Biosciences International 1998. CABI, U.R. (London), pp:99-113.

Migabo, K. (2009). Essai d'introduction de six variétés de riz irrigué dans les conditions edapho climatiques de la plaine de la Ruzizi. Unpublished UCB dissertation: 51 pp.

Misra, R.K. Mathur, S.C. & Chaudhary, R.C. (2016). Breeding rice varieties resistant to diseases. In National Symposium on Increasing Rice Yield in Kharif, pp. 95-114. Cuttack, India, Central Rice Research Institute.

Modibo Ahmadou. (2011). History of Maroua and its neighbourhoods. Manuscript in Arabic (3 p.)
Monty P. Jones. (2016). The rice plant and its environment. ADRA0 training guide, Bouaké (Cote d'Ivoire), n034: 28 p.

Moormann, F.R. (2016). General assessment of land on which rice is grown in West Africa. Paper presented at the Seminar on Rice Soil Fertility and Fertilizer Use, 22-27 January 2016. Monrovia, Liberia, West Africa Rice Development Association.

Mukhopadhya, S.K., Ghosh B.S. & Maity, H. (2014). Weed problem in upland rice and approaches to solve the problem by use of new herbicides. Oryza, 8: 269-274.

Nanda, N. & Prakash Rao, P.S. (2014). Management of insect pests of upland rice. Paper presented at the National Symposium on Multidisciplinary Group Meeting on Rainfed Rice, 17-20 September 1980. Cuttack, India, Central Rice Research Institute.

Nacro, S. (2017). Etude de la bioécologie de la Cécidornyie du riz Orseolia oryzivora sp. n., (Diptera, Cécidomyiidae).et de deux méthodes de lutte contre ce ravageur sur la plaine rizicole de Karfiguéla - (Banfora). Final dissertation, IDR Université de Ouagadougou. 70 p

Nacro S. (2014). Analysis of a tritrophic system: Insect pests of rice and their parasitoids in Cameroon. PhD thesis from the University of Rennes 1 (France), 118 p.

Niyondagara, T. (1984). Contribution à la sélection du riz pour un environnement pluvial. Dissertation, FACAGRO, U.B, Bujumbura: 112pp.
Noëlle Terpend (2017). A practical guide to the value chain approach. The case food supply and distribution in cities. Collection "aliments dans les villes", 39 pages.

Nwilene F. E. (2000). Highlights of WARDA Activities. WARDA Annual Report 2000, WARDA, Bouaké. 30 p.

Okocha, Peter, Markn & Ravindra C. Joshi (2017). Field screening for resistance African rice necidornia. IITA research guide. Ibadan (Nigeria), 16 p.
O'leske, D. L., Robel R. J., & Kemp K. E. (2007). Sweep net-collected invertebrate biomass from high and low-input agricultural fields in Kansas. Wildlife Society Bulletin 25: 133-138.

Ouedraogo, E. & Gouba, N. A. (2003). The use of neem for natural crop protection in the Sahel. Centre Ecologique Albert Schweitzer (CEAS). Ouagadougou (Burkina Faso).

Paba Sale M. (2014). Maroua: aspects de la croissance d'une ville du Nord-Cameroun (des années 50 à nos jours). Univ. Bordeaux-III, these IIIe cycle, 304 p.

Paul Ondo Ovono, Thaddée Gatarasi, Daniel Obame Minko, Drice Miyoumbi Koumagoye & Claire Kevers (2014). Study of insect population dynamics on NERICA rice crop in Masuku conditions, South-East Gabon (Franceville). Int. J. Biol. Chem. Sci. 8 (1): 218-236, February 2014. ISSN 1997-342X (Online), ISSN 1991-8631 (Print).

Poe G. (2015). The single species-single gear bio-economic model: applications to

management and development proposals for the fishery at lake Maga, Pouss. 21 p.
Polaszek, A. & Delvare, G. 2000. Les foreurs des tiges de céréales en Afrique, Cahiers d'Etudes et recherches francophones d'Agriculture, 123pp.
Ponnamperuma, F.N. (2015). Micronutrient limitations in acid tropical rice soils. In E. Borvenusza and A. Alvarado, eds. Soil Management in tropical America, pp. 330-347. Raleigh, North Carolina State University.

Pontie G. (2011). The Guiziga of northern Cameroon. Paris, Mem. Orstom n° 65, 255 p.
Prestat G. (2010). Maroua " ville d'islam ". CHEAM, reference 2, 176, 1953 D, 21 p. dactyl.
Puri H.S. (1999). Neem, the divine tree, Azadirachta indica. 1079 LH Amsterdam.

Qi B., Gordh G., and Walter G. (2001). Effects of neem-fed prey on the predacious insects Harmonia conjormis (Boisduval) (Coleptera: Coccinellidae) and Mallada signatus (Scheneider) (Neuroptera: Chrysopidae). Biological Control, 22: 185-190.

RAI, P.S. & Vidyachandra, N. (2013). Incidence of rice gall midge (Orseolia oryzae (Wood - Mason) in relation to date planting. Regional Research Station, 8 (4): 65- 66.

Rao, M.V. (2016). Pushing up rice yields in uplands. Intensive Agric, 16: 4-7.
Rao, C.N. & Murty, K.S. (2015). Effect of split application of nitrogen on tiller survival and yield in direct seeded upland rice. Indian J. Agric. Sci. 45: 183-185.

Rao, N.V., Rao V.L.V.P., Rao R.H.K., Reddy P.S. (2015). Note on the effect of cultural practices on the incidence of insect pest of rice. Indian Journal of Agriculture Sciences, 52 (l):48 - 50.

Reckhaus P.M. & Adamou I. (1986). Rice diseases and their economic importance in Niger. FAO Plant Protection Bulletin 34 : 77 - 82.

Regnault-Roger C. (2017). Allelochemical molecules and plant extracts in plant protection: nature, role and assessment of their use in the twentieth century. In Regnault-Roger C. phytosanitary issues for agriculture and the environment, Editions Tec & Doc - Lavoisier, Paris, 625-650.

Regnault-Roger C. (2018). Search for new insecticidal biopesticides of plant origin: methodological approach and application to Mediterranean aromatic plants. In Biopesticides of plant origin, 2nd edition, Edition Tec & Doc, pp 25- 49.

Rosenfeld M.F. (2017). Estimating crop yields using the sample cut method. In : Économie rurale. N°18. pp. 25-28.

Roupsard M. (2015). North Cameroon, openness and development. Coutances, Impr. C. Bellee, 516 p.

Roy, J.K., Israel, P. & Panwar, M.S. (2016). Breeding for insect resistance for rice. Oryza, 6: 38-44.

Royer, L. & Boivin G. (2018). Infochemicals mediating the foraging behavior of Aleochara bilineata Gyllenhal adults: sources of attractants. Entomologia Experimentalis et Applicata

90: 199-205.

Sadou I., Woin N., Ghogomu T.R. and Hamasselbe A. (2013). Inventory of predators and parasitoids of insect pests of rice in the far north region of Cameroon. GBS Publishers & Distributors (India), Indian Journal of Applied Agriculture Research. http: //www. gbspublisher. com/Volume/ijaarv1n1.htm.

Sadou I., Woin N., Ghogomu T.R. and Hamasselbe A. (2013). The inventory of insect pests and vectors of rice yellow mottle virus in rice ecosystems in the region of the Far North of Cameroon. GBS Publishers & Distributors (India), Indian Journal of Applied Agriculture Research. http://www.gbspublisher.com/Volume/ijaarv1n1.htm.

Sadou I. (2011). Inventory of insect vectors of rice yellow mottle virus rice ecosystems in the Far North of Cameroon. Book. Editions universitaires européennes, 2011, 84 p. ISBN: 978-613-1-58823-5. Schaltungsdienst Lange, Berlin, Germany. www.thebookedition.com /inventaire - des - insectes - vecteurs - sadou- ismael - p-66593. html or www.ecochimp.com/ inventaire-des-insectes-vecteurs-du-virus-de-la- panachure-jaune-du-riz-sadou-ismael--p-36750.html.

Sadou I., Woin N., Ghogomu T.R. & Djonmaila (2008). Inventory of insect pests and vectors of rice yellow mottle in the irrigated areas of Maga (Far North). Tropicultura, Vol. 26, No. 2, 2008, pp. 84-88.

Sanco D.G. (2011). Review report for the active substance lambda-cyhalothrin Finalised in the Standing Committee on Plant Health at its meeting on 19 October 2000 in view of the inclusion of lambda-cyhalothrin in Annex I of Directive 91/414/eec, european commission directorate general health & consumer protection.

Soraja, R, Raju N. (2016). Effect of nitrogen fertiliser levels and spacing on rice gall midge and leaf - folder damage. InternationalRice Research News/errer 7 (4): 10.

Savary, S., Srivastava R. K., Singh H.M. & Elazegui F.A., (2017). A characterization of rice pests and quantification of yield Josses in the rice-wheat system of india. Crop. Prat. 16 (4) p : 387-398.

Sawadogo, A. (2017). Pest knowledge and crop defence. Notion de nématologie agricoles. Institut d'Etudes el de Recherches Agricoles Training workshop for lNERA technicians, Ouagadougou, 12 p.

Schalbroeck, J.J. (2001). Rice: 85-105 In: Raemakers, R.JH. (Ed): Agriculture en Afrique Tropical. D.G.C.I., Rue de petits carmes, Brussels, Belgium :1634PP.

Schalbroeck J.J. (2001). Cereal plants. Agriculture in Tropical Africa 1: 10-169.

Schalbroeck J. J. (2016). Rice. In: Agriculture in Tropical Africa. Ministry of Foreign Affairs of Foreign Trade and International Cooperation, Brussels, Belgium, p. 99 - 105.

Schmutterer, H. (1990). Properties and potential of natural pesticides from the neem tree, Azadirachta indica. Annual of Review Entomology 35: 271-297.

Sebillotte M. (2016). Agronomy and agriculture. Essai d'analyse des taches de l'agronome,

Cah. ORSTOM, ser. Biol, (24).

Seignobos C. (2015). L'arbre et la cite dans la zone soudano-sahélienne (les exemples du Tchad et du Cameroun septentrional). Yaounde, Rev. de Géogr. du Cameroun, 2 (1): 49-52.

Sere, Y. (2014). The control of rice blast in Upper Volta. Proceedings of the Symposium on rice resistance to blast. Montpellier - France. pp. 51-66.

Sere, Y. (2015). Rice blast in irrigated rice cultivation. Example of the kou valley (Burkina Faso). Phytosanitary problems of rice in Burkina Faso and review of research. Sahel info July 2015, pp: 5-6.

Sere, Y. & Eble S. (2017). Rice production and rice research in Burkina Faso. Report presented at the session of the international rice commission held in GOIANA, Brazil, 4-9 February 1990. 25 p.

Shaaya E, Ravid U, Paster N, Juven B, Lisman U & Pissarev V. (2016). Fumigant toxicity of essential oils against four majors stored-products insects. J Chem Ecol, 173p (pp499-504).

Shastry, S.V.S., Sharma, S.D., John, V.T. & Krishnaiah, K. (2016). New sources of resistance to pests and diseases in Assam rice collection. Int. Rice Comm Newsl. 20 (2): 1-16.

Sie M., Kabore K. B., Dakouo D., Dembele Y., Sedga Z., Bado B. V., Ouedraogo M., Moukoumbi Y. D., Ban. M. & A. Traore. (2006). Fiches techniques des quatre nouvelles variétés de riz de type NERICA pour la riziculture de bas-fond / irriguée au Burkina Faso : FKR 56 N, FKR 58 N, FKR 60 N, FKR 62 N. INERA, Bobo Dioulasso. Burkina Faso, 5 p.

Silvestre A. (2014). "The diffusion of maize in North Cameroon: innovation dynamics and local technical culture". Thesis in geography from the école des hautes études en sciences sociales, directed by Boutrais J, Paris.

Singh B., Sharma D.K., Kumar R., Gupta A. (2010). Development of a new controlled pesticide delivery system based on neem leaf powder. Journal of Hazardous Materials 177 : 290-299.

Singh, R.K. & Singh, C.V. (2016). Central Rainfed Upland Rice Research Station annual report 2016. Hazaribagh, India, ICAR.

Singh, G.R, Parihar S.S. & Chaure (2017). Response of organic manures in a rice (Oryza sativa) - Chickpea (Cicer arietinum) crop sequence. Taxonomy of the genus Oryza (Poaceae): International Rice Research Newsletter 24 (3): 24 - 25.

Stoltz D.B. (1986). Interactions between parasitoid derived product and host insects: an over view. Journal of insect physiology. 32: 347-357.

Strand, M.R. (2016). The physiological interactions of parasitoids with their hosts and their influence on reproductive strategies. Pages 97-136 in E. Wajnberg and D. Greathead (eds.), Insect parasitoid. Academic Press, London.

Sukmana, S. & Suwardjo, H. (2017). Prospects of vegetative soil conservation measures for sustainable upland agriculture. IARD J., 13 (1-2): 1-7.

Sullivan, D.J. (2017). Insect hyperparasitism. Annual review of entomology 32: 49-70.
Sunderland K.D. & Chanbers R.J. (1983). Invertebrate polyphagous predators as post control agents. Some criteria and methods. Pp 100-108 In: Cavalloro, Aphid antagonists. Balkema, Rotterdam.

Sy A A. (1994). Special Rice Yellow Mottle Virus monitoring to Mali. WARDA, Bouaké, Côte d'Ivoire. 28 p.

Sy A.A. & Alluri K. (1993). Selection of RYMV resistant and estimation of RYMV induced yied loss under natural and artificial pressure. Pp 44-45 In: WARDA, 1993. Annual Report, WARDA, Bouaké.

Swzedo, J., Th. Bourgoin & F. Lefèbvre. (2004). An annotated catalogue with notes on Hemiptera classification.

Tagne, A. Nguefack, J. The, C. Amvam, Z.H. (2002). Introduction to bio-pesticides: botanical pesticides. Bulletin d'Information Phytosanitaire 33: 23.

Tamgno, B.R. 2011. Diagnostic des problems agricoles en champs comme en stockage dans la Vallée du Logone. Rapport d'activites, PRASAC, Ngaoundere. 20 p.
Tirard C., R. Barbault, L. Abbadie, N. Loeuille (2012). Mini manuel d'écologie, cours+QCM/QROC, Dunod, p. 33-37.

To Bi T. J-L. (2016). Contribution à l'étude de l'efficacité des extraits végétaux en culture maraîchères : cas Azadirachta indica Juss et de Carica papaya L. Final dissertation for the Diplôme d'Ingénieur des Techniques Agricoles, Ecole supérieure d'Agronomie de Yamoussoukro, 66 p.

Tougouma, A. (2014). Contribution to the study of the effect of environmental factors on host-parasite relationships in the rice-Pyricularia Oryzae cav pair. Final dissertation. University of Ouagadougou, 72 p

Toure, K. (2017). Biology and ecology of Chilo zaccouius Blesz, rice stem borer in Mali. Tropical relationships with its cultivated host plant: rice. Sahel Pl' info, pp: 4-7.

Tran, D.V. (2016). Overview of upland rice in the world. In Progress in Upland Rice Research, pp. 51-66. Los Baños, Philippines, IRRI.

United States Department of Agriculture (USDA). (2014). National agricultural statistics service agricultural statistics 2014, united states government printing office washington: 2014

Van Driesche, R.G. & Bellows T.S. (2016). Biological control. Chapman & Hall. Toronto.
Vardhani, B.P. & K.T. Rao (2002). Effect of normal and sub-normal application of insecticides on brown planthopper and its predators. Journal of Applied Zoological Researches 13:74-75.

Vaughan, D. A., Morishima, H. & Kadowaki, K. (2003). Diversity in the Oryza genus. Current Opinion in Plant Biology, 6: 139-146.
Vermura, Y. & Mayasaka, A. (2018). Causes of rice yield decrease resulting from continual

direct sowing culture and its relation to phosphorus nutrition. Proc. Crop Sci. Soc. Japan, 42 : 116-122.

Vincent J.F. (2016). ≪ Donnees nouvelles sur la fondation de la chefferie de Marva et le peuplement Giziga-Mofu de la region (Nord-Cameroun)≫ . Colloque CNRS, L'esprit de découverte, Valbonne, dec. 2016, 24 p. multigr.

Vinson S.B. & Iwantsch GF (1980). Host for insect parasitoids. Annual Review of Entomology 25: 397-419.

Vinson, S.B. (2014). Physiological interactions between egg parasitoids and their hosts. Pages 201-218 in E. Wajnberg and S.A. Hassan (eds.), Biological control with egg parasitoids. Cab International, Wallingford.

WARDA (2006). Progress Report 2003-2005 joint interspecific hybridization between African and Asian Rice species (Oryza glaberrima and O. sativa), P. 151. Cotonou, Benin.

Waage J. & Greathead D. (1986). Insect parasitoids. Pp 388-389 In: Academic press, London.

Woïn N., Volkmar C. & Wetzel T. (2004). Seasonal activity of ladybirds (Coleoptera: Coccinellidae) as ecological bioindicators in paddy fields. Deutsch Journal of Entomology 12: 203-206.

Woin N., Sadou I., Djomaila (2006). Inventory and ecology of insect vectors of rice yellow mottle disease and their potential natural enemies in the rice-growing ecosystems of Cameroon. Abstract Congrès du riz en Afrique : Au-delà de la première génération des NERICA en Afrique. 31 July-4 August 2006. Dar es Salaam, Tanzania. Journal abstracts rice congress in Africa, 2006, p. 75.

Woin N., Sadou I., Djomaila (2006). Performance of some NERICA rice varieties under the pressure of entomofauna pests. Abstract Congrès du riz en Afrique : Au-delà de la première génération des NERICA en Afrique. 31 July-4 August 2006. Dar es Salaam, Tanzania. Journal abstracts rice congress in Africa, 2006, p. 59.

Woin, N. ; Ismael, Sadou ; Bourou, Sali & Bebom, T. (2007). Potential for Biological Control of Rice Yellow Mottle Virus Vectors. African Crop Science Journal, Vol. 15, No. 4, 2007, pp. 211-222.

Yao N. D. (2010). Study of the efficacy of neem (Azadirachta indica Juss) leaf and seed extracts on okra (Abelmoschus esculentus L.) insects: Flea beetles, caterpillars and Chinch bugs. Final dissertation for the Agronomy Engineering diploma, EFCPC of INP-HB (Yamoussoukro), 60p.

Young, O.P. & T.C. Lockley (2015). The striped lynx spider, Oxyopes salticus (Araneae: Oxyopidae), in agroecosystems. Entomophaga 30 :329-346.

Zadia, H, C. (2016). Analyse de la compétitive dans la gestion de la diversité variétale de riz dans la plaine de la Ruzizi (cas de lubarika, luberizi, et kiringye), unpublished dissertation UCB: 40pp.

Zhang, Y. (2004). Factors affecting chines farms decision to adapt a water saving technology.

Can.j. agric. econ. 56, 61pp.

Zongo, O. J., Vincent, C. & Steward, R K. (2016). Effectiveness of four types of traps for the capture of Atherigona soccata Rondani (Diptera: Muscidae) and effects of some cultural practices on this damage. International seminar on integrated pest management. Pp : 116- 126.

APPENDICES

Appendix 1: Some photos of technical operations in the field

Photo: The experimental plots

Photo: Collecting insects

Photo: spraying the biopesticide

Photo: insects caught

Photo: setting up the experimental rice parcel

APPENDIX 2: Photograph of certain species insect pests of rice

Hydrellia philippina Ferino

(Diptera: Ephydridae)

Nymphula depunctalis (Guenee)

(Lepidoptera: Pyralidae)

Rivula atimeta Swinhoe

(Lepidoptera: Noctuidae)

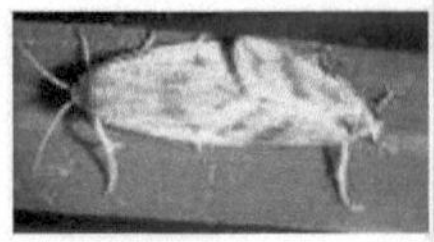

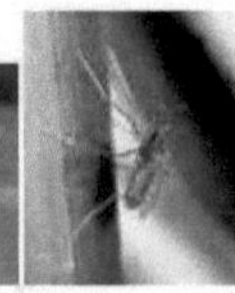

Rivula atimeta (Swinhoe)	*Naranga aenescens* Moore	*Orseolia oryzae*
(Lepidoptera: Noctuidae)	(Lepidoptera: Noctuidae)	(Diptera: Cecidomyiidae)

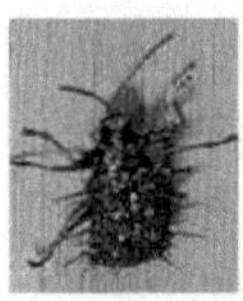

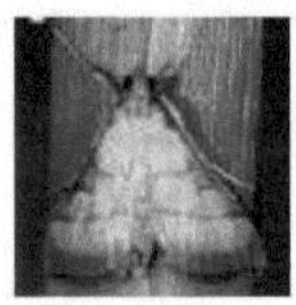

Dicladispa armigera (Olivier)	*Cnaphalocrocis medinalis*	*Marasmia exigua* Butler
(Coleoptera: Chrysomelidae)	(Lepidoptera: Pyralidae)	(Lepidoptera: Pyralidae)

Marasmia patnalis Bradley

(Lepidoptera: Pyralidae)

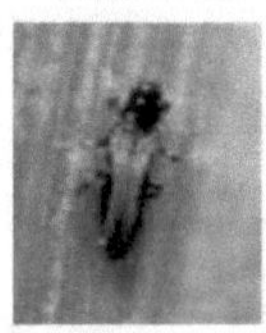

Stenchaetothrips biformis Bagnall

(Thysanoptera: Thripidae)

Pelopidas mathias Fabricius

(Lepidoptera: Hesperiidae)

Parnara guttata Bremer

(Lepidoptera: Hesperiidae)

Oxya sp.

(Orthoptera: Acrididae)

Acrida sp.

(Orthoptera: Acrididae)

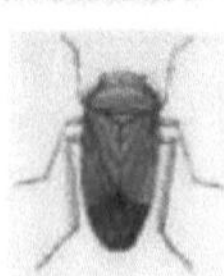

Nephotettix cincticeps

(Hemiptera: Cicadellidae)

Nephotettix nigropictus

(Hemiptera: Cicadellidae)

Nephotettix virescens

(Hemiptera: Cicadellidae)

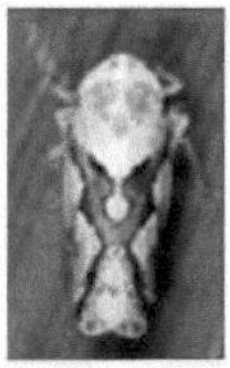

Recilia dorsalis (Motschulsky)

(Hemiptera: Cicadellidae)

Nilaparvata lugens (Stal)

(Hemiptera: Delphacidae)

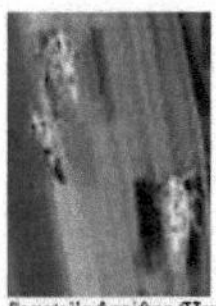

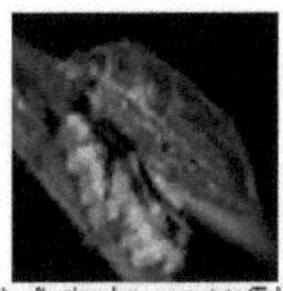

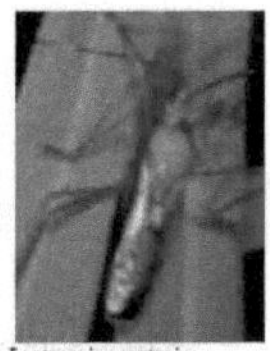

Sogatella furcifera (Horvath)

(Hemiptera: Delphacidae)

Scotinophara coarctata (Fabricius)

(Hemiptera: Pentatomidae)

Leptocorisa oratorius

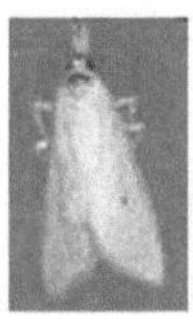

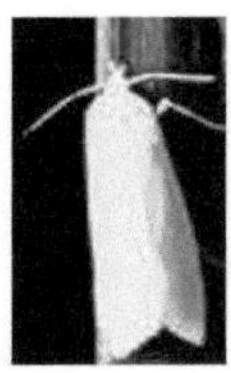

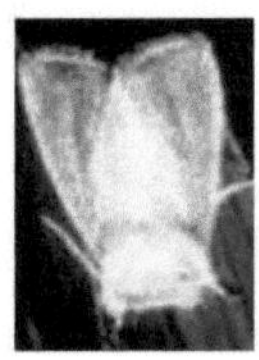

Scirpophaga incertulas (Walker)	*Scirpophaga innotata* (Walker))	*Sesamia inferens* (Walker)
(Lepidoptera: Pyralidae)	(Lepidoptera: Pyralidae)	(Lepidoptera: Noctuidae)

Chilo suppressalis (Walker)

(Lepidoptera: Pyralidae)

Chilo auricilius Dudgeon

(Lepidoptera: Pyralidae)

Chilo polychrysus (Meyrick)

(Lepidoptera: Pyralidae)

Diopsis thoracica West Wood

(Diptera: Diopsidae)

FSC
www.fsc.org

MIX
Papier aus verantwortungsvollen Quellen
Paper from responsible sources
FSC® C105338

Printed by Books on Demand GmbH, Norderstedt / Germany